Remix

Also by Lawrence Lessig

Code and Other Laws of Cyberspace

The Future of Ideas

Free Culture

Code: Version 2.0

Remix

Making **art** and
commerce thrive
in the hybrid
economy

LAWRENCE LESSIG

BLOOMSBURY

First published in the United States by The Penguin Press, 2008

First published in Great Britain 2008

Bloomsbury Academic
An imprint of Bloomsbury Publishing plc
36 Soho Square
London W1D 3QY
www.bloomsburyacademic.com

A CIP record for this book is available from the British Library.

ISBN: 9-781-4081-1347-9

This book is produced using paper that is made from wood grown in managed, sustainable forests.
It is natural, renewable and recyclable. The logging and manufacturing processes conform to the
environmental regulations of the country of origin.

Printed and bound in Great Britain by Clays Ltd, St Ives plc

To two teachers,

L. Ray Patterson and Jack Valenti

CONTENTS

PREFACE

In early 2007, I was at dinner with some friends in Berlin. We were talking about global warming. After an increasingly intense exchange about the threats from climate change, one overeager American at the table blurted, "We need to wage a war on carbon. Governments need to mobilize. Get our troops on the march!" Then he fell back into his chair, proud of his bold resolve, sipping a bit too much of the wildly too-expensive red wine.

It was obvious that my friend was speaking metaphorically. Carbon is not an "enemy." Not even an American marine could fight it. Yet, as I looked around the table, a kind of reticence seemed to float above our German companions. "What does that look mean?" I asked one of my friends. After a short pause, he almost whispered, "Germans don't like war."

The response sparked a rare moment of recognition (in me). Of course, no one was talking about using guns to fight carbon. Or even carbon polluters. Yet, for obvious reasons, the associations with war in Germany are strongly negative. The whole country, but especially Berlin, is draped in constant reminders of the costs of that country's twentieth-century double blunder.

But in America, associations with war are not necessarily

negative. I don't mean that we are a war-loving people; I mean that
our history has allowed us to like the idea of waging war. Not out
of choice, but as a remedy to a great wrong. War is a sacrifice that
we have made, and in one recent case at least, a sacrifice to a very
good end. We thus romanticize that sacrifice.

That romance in turn allows the metaphor to spread into other
social or political conflicts. We wage war on drugs, on poverty, on
terrorism, on racism. There is a war on government waste, a war on
crime, a war on spam, a war on guns, and a war on cancer. As Professors George Lakoff and Mark Johnson describe, each of these
"wars" produces a "network of entailments." Those entailments
then frame and drive social policy. As they put it, in discussing
President Carter's "moral equivalent of war" speech:

> There was an "enemy," a "threat to national security," which
> required "setting targets," "reorganizing priorities," "establishing
> a new chain of command," "plotting new strategy," "gathering
> intelligence," "marshaling forces," "imposing sanctions," "calling
> for sacrifices," and on and on. The WAR metaphor highlighted
> certain realities and hid others. The metaphor was not merely a
> way of viewing reality; it constituted a license for policy change
> and political and economic action. The very acceptance of the
> metaphor provided grounds for certain interferences: there was
> an external, foreign, hostile enemy (pictured by cartoonist in Arab
> headdress); energy needed to be given top priorities; the populace
> would have to make sacrifices; if we didn't meet the threat we
> would not survive.[1]

A fight for survival has obvious implications. Such fights get
waged without limit. It is cowardly to question the cause. Dissent is

an aid to the enemy—treason, or close enough. Victory is the only result one may contemplate, at least out loud. Compromise is always defeat.

These entailments make obvious sense during conflicts such as World War II, when there really was a fight for survival; my spark of Lakoffian recognition, however, was to see just how dangerous these entailments are when the war metaphor gets applied in contexts in which, in fact, survival is not at stake.

Think, for example, about the "war on drugs." Fighting debilitating chemical addiction is no doubt an important social objective. I have seen firsthand the absolute destruction it causes. But the "war on drugs" metaphor prevents us from recognizing that there may be other, more important objectives that the war is threatening. Think about the astonishingly long prison terms facing even small-time dealers—the Supreme Court, for example, has upheld a life sentence without the possibility of parole for the possession of 672 grams of cocaine.[2] Think about ghettos burdened by the drug trade. Think about governments in Latin America that have no effectively independent judiciary or even army because the wealth produced by prohibition enables the drug lords to capture their control. And then think about the fact that this war has had essentially no effect on terminating the supply of drugs. One doesn't notice these inconvenient truths in the middle of a war. To see them, you need a truce. You need to step back from the war to ask, How much is it really costing? Is the results really worth the price?

The inspiration for this book is the copyright wars, by which right-thinking sorts mean not the "war" on copyright "waged" by "pirates" but the "war" on "piracy," which "threatens" the "survival" of certain important American industries.

This war too has an important objective. Copyright is, in my view at least, critically important to a healthy culture. Properly balanced, it is essential to inspiring certain forms of creativity. Without it, we would have a much poorer culture. With it, at least properly balanced, we create the incentives to produce great new works that otherwise would not be produced.

But, like all metaphoric wars, the copyright wars are not actual conflicts of survival. Or at least, they are not conflicts for survival of a people or a society, even if they are wars of survival for certain businesses or, more accurately, business models. Thus we must keep in mind the other values or objectives that might also be affected by this war. We must make sure this war doesn't cost more than it is worth. We must be sure it is winnable, or winnable at a price we're willing to pay.

I believe we should not be waging this war. I believe so not because I think copyright is unimportant. Instead, I believe in peace because the costs of this war wildly exceed any benefit, at least when you consider changes to the current regime of copyright that could end this war while promising artists and authors the protection that any copyright system is intended to provide.

In the past, I've tried to advance this view for peace by focusing on the costs of this war to innovation, to creativity, and, ultimately, to freedom. My aim in *The Future of Ideas* was to defend industries that never get born for fear of the insane liability that the current regime of copyright imposes. My subject in *Free Culture* was the forms of creative expression and freedom that get trampled by the extremism of defending a regime of copyright built for a radically different technological age.

But I finished *Free Culture* just as my first child was born. And in the four years since, my focus, or fears, about this war have changed. I don't doubt the concerns I had about innovation, creativity, and freedom. But they don't keep me awake anymore. Now I worry about the effect this war is having upon our kids. What is this war doing to them? Whom is it making them? How is it changing how they think about normal, right-thinking behavior? What does it mean to a society when a whole generation is raised as criminals?

This is not a new question. Indeed, it was the question that the former, now late, head of the Motion Picture Association of America, Jack Valenti, asked again and again as he fought what he called a "terrorist war" against "piracy."[3] It was the question he asked a Harvard audience the first time he and I debated the issue. In his brilliant and engaging opening, Valenti described another talk he had just given at Stanford, at which 90 percent of the students confessed to illegally downloading music from Napster. He asked a student to defend this "stealing." The student's response was simple: Yes, this might be stealing, but everyone does it. How could it be wrong? Valenti then asked his Stanford hosts: What are you teaching these kids? "What kind of moral platform will sustain this young man in his later life?"

This wasn't the question that interested me in that debate. I blathered on about the framers of our Constitution, about incentives, and about limiting monopolies. But Valenti's question is precisely the question that interests me now: "What kind of moral platform will sustain this young man in his later life?" For me, "this young man" represents my two young sons. For you, it may be your daughter, or your nephew. But for all of us, whether we have kids or not, Valenti's question is exactly the question that should concern us most. In a world in which technology begs all of us to create and spread creative work differently from how it was created and spread before, what kind of moral platform will sustain our kids, when their ordinary behavior is deemed criminal? Who will they become? What other crimes will to them seem natural?

Valenti asked this question to motivate Congress—and anyone else who would listen—to wage an ever more effective war against "piracy." I ask this question to motivate anyone who will listen (and Congress is certainly not in that category) to think about a different question: What should we do if this war against "piracy" as we currently conceive of it cannot be won? What should we do if we know that the future will be one where our kids, and their kids, will use a digital network to access whatever content they want whenever they want it? What should we do if we know that the future is one where perfect control over the distribution of "copies" simply will not exist?

In that world, should we continue our ritual sacrifice of some kid caught downloading content? Should we continue the expulsions from universities? The threat of multimillion-dollar civil judgments? Should we increase the vigor with which we wage war against these "terrorists"? Should we sacrifice ten or a hundred to a federal prison (for their actions under current law are felonies), so that others learn to stop what today they do with ever-increasing frequency?

In my view, the solution to an unwinnable war is not to wage war more vigorously. At least when the war is not about survival, the solution to an unwinnable war is to sue for peace, and then to find ways to achieve without war the ends that the war sought. Criminalizing an entire generation is too high a price to pay for almost any end. It is certainly too high a price to pay for a copyright system crafted more than a generation ago.

This war is especially pointless because there are peaceful means to attain all of its objectives—or at least, all of the legitimate objectives. Artists and authors need incentives to create. We can craft a system that does exactly that without criminalizing our kids. The last decade is filled with extraordinarily good work by some of the very best scholars in America, mapping and sketching alternatives to the existing system. These alternatives would achieve the same ends that copyright seeks, without making felons of those who naturally do what new technologies encourage them to do.

It is time we take seriously these alternatives. It is time we stop wasting the resources of our federal courts, our police, and our universities to punish behavior that we need not punish. It is time we stop developing tools that do nothing more than break the extraordinary connectivity and efficiency of this network. It is time we call a truce, and figure a better way. And a better way means redefining the system of law we call copyright so that ordinary, normal behavior is not called criminal.

Many will read this declaration and wonder just why I should be allowed to teach law at a great American university. Do we respond to high levels of rape by decriminalizing rape? Would tax evasion best be solved by eliminating taxes? Should the fact of speeding mean we should repeal the speed limit? Or put generally: Does the fact of crime justify the repeal of criminal law?

Of course not. Rape is wrong and should be punished severely whether or not people continue to rape. Tax evasion is evil and should be punished much more severely than it is, whether or not most people cheat. And speeding kills and should be regulated much more effectively than it is now, even if most of us regularly speed. Nothing I'm saying about the copyright war in particular generalizes automatically to every other area of regulation. I am talking specifically about one unwinnable war, and about alternatives to that war that have the consequence of decriminalizing our kids, and decriminalizing many of us too.

But I confess that I do believe that this way of thinking about the copyright wars should affect how we think about other kinds of regulation. Tax evasion is wrong. But one way to avoid that wrong would be a simpler, fairer tax system. Speeding is wrong. But one way to avoid that wrong is to avoid fifty-five-mile-per-hour speed limits on straight, rural, four-lane public highways. We should always be thinking about how to moderate regulation in light of the likelihood that the target of regulation will comply. It does no one any good to regulate in ways that we know people will not obey.

We need, in other words, more humility about regulation. The twentieth century changed us in many obvious ways. But the one way we're likely not to notice is the presumption the twentieth century gave us that government regulation is plausibly successful. For most of the history of modern government, the struggle was not about what was good or bad; the struggle was about whether it was possible to imagine government effecting any good through regulation. Fears of inevitable corruption, in part at least, drove our framers to limit the size of the federal government—not idealism about libertarianism. Recognizing the uselessness of certain sorts of rules led governments to avoid regulation in obvious areas, or to deregulate when they saw their regulation failing. These are the historical expressions of regulatory humility, a habit of mind for most of human history.

We've forgotten these limits of humility. Wherever there is a wrong, the first instinct of our government is to send in the legal equivalent of the marines. We pass a law to ban a behavior, but we rarely work through just how that law will change behavior. Nor do we assess how corrosive it is if, the law notwithstanding, the behavior remains the same, though now with the label "criminal." If something is wrong, it gets a law, without us even working through alternatives to exploding regulation.

If you're skeptical, think about a simple example. Around the time the Supreme Court heard arguments in the well-known peer-to-peer file-sharing case *MGM v. Grokster*,[4] my local public-radio station aired a story about the case. The story happened to run on a day when the radio station was also running its own fund-raising drive. Just after the story about *Grokster* ended, the show shifted to its call for public support. "More than 90 percent of people who listen to public radio don't contribute to its support," the announcer said. "That's why we need you to contribute now."

I had worked on a brief in the *Grokster* case, in which we addressed the content industry's claim that 91 percent of the content shared on peer-to-peer file-sharing networks was in violation of copyright law. We had responded by reminding the Court that in the earlier Sony Betamax case, in which the VCR was the target, the content industry had also estimated that 91 percent of VCR usage was in violation of copyright laws. The industry was nothing if not consistent.

But the contrast between the complaint in the Supreme Court and the complaint of the announcer on public radio startled me. Here were two examples of free riding: people downloading Britney Spears's music without paying her and people listening to "All Things Considered" without paying NPR. With one, we criminalize the free riding. With the other, we don't. Why? Do you think it would be appropriate to arrest people who listen to NPR without paying? I certainly don't. And as you may wonder, do I think Britney Spears should be paid through voluntary pledges on a 1-800 number? No, again, I don't.

My point in retelling this story is to get you to see something that is otherwise too often obscure: there are many different ways in which we tax to raise the revenues needed for public goods (as the economist would call copyrighted works). We select among these different ways the one that is best. The critical point I want this book to make is that one factor we should consider when deciding that is whether the way we select makes our kids criminals. That's not the only factor. But it is one that has plainly been missing from Congress's consideration about how best to deal with the impact of digital technologies upon traditional copyright industries.

REMIX

INTRODUCTION

In early February 2007, Stephanie Lenz's eighteen-month-old son, Holden, started dancing. Pushing a walker across her kitchen floor, Holden started moving to the distinctive beat of a song by Prince (that's the current name of the artist formerly known as Prince), "Let's Go Crazy." Holden had heard the song a couple of weeks before while the family watched the Super Bowl. The beat had obviously stuck. So when he heard the song again, he did what any sensible eighteen-month-old would do—he accepted Prince's invitation and went "crazy" to the beat, in the clumsy but insanely cute way that any precocious eighteen-month-old would.

Holden's mom, understandably, thought the scene hilarious. She grabbed her camcorder and captured the dance digitally. For twenty-nine seconds, she had the priceless image of Holden dancing, with the barely discernible Prince playing on a radio somewhere in the background.

Lenz wanted her parents to see the film. But it's a bit hard to e-mail a 20-megabyte video file to anyone, including your relatives. So she did what any sensible citizen of the twenty-first century would do: she uploaded the file to YouTube and e-mailed her relatives the link. They watched the video scores of times, no doubt sharing the link with friends and colleagues at work. It was a perfect YouTube moment: a community of laughs around a homemade video, readily shared with anyone who wanted to watch.

Sometime over the next four months, however, someone not a friend of Stephanie Lenz also watched Holden dance. That someone worked for Universal Music Group. Universal either owns or administers some of the copyrights of Prince. And Universal has a long history of aggressively defending the copyrights of its authors. In 1976, it was one of the lead plaintiffs suing Sony for the "pirate technology" now known as the VCR. In 2000, it was one of about ten companies suing Eric Corely and his magazine, *2600,* for publishing a link to a site that contained code that could enable someone to play a DVD on Linux. And now, in 2007, Universal would continue its crusade against copyright piracy by threatening Stephanie Lenz. It fired off a letter to YouTube demanding that it remove the unauthorized performance of Prince's music. YouTube, to avoid liability itself, complied.

This sort of thing happens all the time today. Companies like YouTube are deluged with demands to remove material from their systems. No doubt a significant portion of those demands are fair and justified. If you're Viacom, funding a new television series with high-priced ads, it is perfectly understandable that when a perfect copy of the latest episode is made available on YouTube, you would be keen to have it taken down. Copyright law gives Viacom that power by giving it a quick and inexpensive way to get the YouTubes of the world to help it protect its rights.

The Prince song on Lenz's video, however, was something completely different. First, the quality of the recording was terrible. No one would download Lenz's video to avoid paying Prince for his music. Likewise, neither Prince nor Universal was in the business of selling the right to video-cam your baby dancing to their music. There is no market in licensing music to amateur video. Thus, there was no plausible way in which Prince or Universal was being harmed by Stephanie Lenz's sharing this video of her kid dancing with her family, friends, and whoever else saw it. Some parents might well be terrified by how deeply commercial culture had penetrated the brain of their eighteen-month-old. Stephanie Lenz just thought it cute.

Not cute, however, from Lenz's perspective at least, was the notice she received from YouTube that it was removing her video. What had she done wrong? Lenz wondered. What possible rule—assuming, as she did, that the rules regulating culture and her (what we call "copyright") were sensible rules—could her maternal gloating have broken? She pressed that question through a number of channels until it found its way to the Electronic Frontier Foundation (on whose board I sat until the beginning of 2008).

The EFF handles lots of cases like this. The lawyers thought this case would quickly go away. They filed a counternotice, asserting that no rights of Universal or Prince were violated, and that Stephanie Lenz certainly had the right to show her baby dancing. The response was routine. No one expected anything more would come of it.

But something did. The lawyers at Universal were not going to back down. There was a principle at stake here. Ms. Lenz was not permitted to share this bit of captured culture. They would insist—indeed, would threaten her with this claim directly—that sharing this home movie was willful copyright infringement. Under the laws of the United States, Ms. Lenz was risking a $150,000 fine for sharing her home movie.

We'll have plenty of time to consider the particulars of a copyright claim like this in the pages that follow. For now, put those particulars aside. Instead, I want to you imagine the conference room at Universal where the decision was made to threaten Stephanie Lenz with a federal lawsuit. Picture the meeting: four, maybe more, participants. Most of them lawyers, billing hundreds of dollars an hour. All of them wearing thousand-dollar suits, sitting around looking serious, drinking coffee brewed by an assistant, reading a memo drafted by a first-year associate about the various rights that had been violated by the pirate, Stephanie Lenz. After thirty minutes, maybe an hour, the executives come to their solemn decision. A meeting that cost Universal $10,000? $50,000? (when you count the value of the lawyers' time, and the time to prepare the legal materials); a meeting resolved to invoke the laws of Congress against a mother merely giddy with love for her eighteen-month-old.

Picture all that, and then ask yourself: How is it that sensible people, people no doubt educated at some of the best universities and law schools in the country, would come to think it a sane use of corporate resources to threaten the mother of a dancing eighteen-month-old? What is it that allows these lawyers and executives to take a case like this seriously, to believe there's some important social or corporate reason to deploy the federal scheme of regulation called copyright to stop the spread of these images and music? "Let's Go Crazy"? Indeed! What has brought the American legal system to the point that such behavior by a leading corporation is considered anything but "crazy"? Or to put it the other way around, who have we become that such behavior seems sane to anyone?

Near the center of London, in a courtyard named Mason's Yard, there is a modern-looking cement building called White Cube. In a previous life, it was an electricity substation. Today it is an art gallery.

In late August 2007, I entered the gallery and walked to the basement. A large black curtain separated the stairs from an exhibit. When I passed through the curtain, I saw on one wall of the huge black room twenty-five plasma displays, one set next to the other, in portrait orientation. Each display was a window into a studio. In each studio was a fan of John Lennon. Twenty-five fans—three women, twenty-two men, fifteen wearing T-shirts (both men and women), one wearing a tie (man). All twenty-five were singing the vocal track, from the first song to the last, without pause, from John Lennon's first solo album, *John Lennon/Plastic Ono Band* (1970). The exhibit looped the video again and again, for eight hours a day, six days a week, throughout the summer of 2007.

These fans were ordinary Brits. Very ordinary. None were beautiful. None were very young. They had no makeup. They were twenty-five Lennon fanatics, selected from over six hundred who had applied to sing this tribute to their favorite artist.

London was not the only city with an exhibit like this. Three related installations had been made in three different countries. In Jamaica, *Legend (A Portrait of Bob Marley)* featured thirty fans singing Marley's *Legend* album. In Berlin, *King (A Portrait of Michael Jackson)* had sixteen fans singing the whole of *Thriller*. And in Italy, thirty fans of Madonna gathered for *Queen (A Portrait of Madonna)*, a tribute to the queen of pop. *Working Class Hero (A Portrait of John Lennon)* was just the latest in the series. The young South African artist who had created it, Candice Breitz, was considering making more.

I'm not one to be moved by John Lennon's solo work. Yet as I sat in that pitch-black room, watching these fans sing his music, I was overwhelmed with emotion. Like a mother holding her baby for the first time, or a boy reaching out to take his father's hand, or a daughter turning to kiss her father as her wedding begins, each of these fans conveyed an extraordinary and contagious emotion. They were not fantastic singers. Often someone would miss the timing or forget the words. But you could see that this music and its creator were among the most important things in these people's lives. Who knows why? Who knows what their particular associations were? But it was clear that this album was just about the most important creative work these fans knew. Their performance was a celebration of this part of their lives. That was its point: not so much about Lennon, but about the people whose lives Lennon had touched.

Throughout her career Breitz has focused upon the relationship between mainstream culture—from blockbuster movies to pop music—and the audience who experiences it. As she explained to me,

the idea is to shift the focus away from those people who are usually perceived as creators so as to give some space, some room, to those people who absorb cultural products—whether it's music or movies or whatever the case may be. And to think a little bit about what happens once music or a movie has been distributed: how it may get absorbed into the lives into the very being of the people who listen to it or watch it.[1]

Each of us connects differently. The connection runs deep in some; it skips across the surface in others. Sometimes it catches us and pulls us along. Sometimes it changes us completely. Again, Breitz:

> Even the most broadly distributed, most market-inflected music comes to have a very specific and local meaning for people according to where it is that they're hearing it or at what moment in their life they're hearing it. What goes hand in hand with the moment of reception is a dimension of personal translation.

This "reception," she continued, "involves . . . interpretation or translation." That act "is creative." Active. Engaged. Yet, it's easy for us to miss the active in the mere watching. It's rude to turn around and watch people watch a movie. It's a crime to try to film them singing in the shower. We live in a world infused with commercial culture, yet we rarely see how it touches us, and how we process it as it touches us.

As Breitz explained this to me, I wondered about its source in her. Where did it come from? I asked her. In part, it was African.

> In African and other oral cultures, this is how culture has traditionally functioned. In the absence of written culture, stories and histories were shared communally between performers and their audiences, giving rise to version after version, each new version surpassing the last as it incorporated the contributions and feedback of the audience, each new version layered with new details and twists as it was inflected through the collective. This was never thought of as copying or stealing or intellectual-property theft but accepted as the natural way in which culture evolves and

develops and moves forward. As each new layer of interpretation was painted onto the story or the song, it was enriched rather than depleted by those layers.

But this reality is not unique to oral cultures. In Breitz's view, it is "how the artistic process works" generally.

This process of making meaning may be more blatant in the practice of certain artists than it is in the practice of others. Artists who work with found footage, for example, blatantly reflect on the absorptive logic of the creative process. But I would argue that every work of art comes into being through a similar process, no matter how subtly. No artist works in a vacuum. Every artist reflects—consciously or not—on what has come before and what is happening parallel to his or her practice.

This understanding of culture, and the artist's relationship to culture, led directly to the particular work I was watching at White Cube. As she described to me,

these works are based on a pretty simple premise: there are enough images and representations of superstars and celebrities in the world. Rather than creating more images of people who are already overrepresented, rather than literally making another image of a Madonna or a John Lennon, I wanted to reflect on the other side of the equation, on what goes into the making of celebrity.

I realized I needed to turn the camera 180 degrees, away from those who are usually in the public eye—those who already have a strong voice and presence on the screen or stage—towards those on the other side of the screen or stage, the audience members who attend concerts, watch movies, and buy CDs.

Towards those who are usually—incorrectly, in my opinion—conceived of as mere absorbers of culture rather than being recognized as having the potential to reflect culture creatively.

Prior to *Working Class Hero*, the similar installations had all been well received. After seeing *Legend*, for example, Bob Marley's widow, Rita, decided to incorporate permanently a copy in the inventory of the Bob Marley Museum in Kingston, where she had arranged an opening showing at the museum, inviting all thirty performers and their families from across Jamaica to come to the museum to celebrate its celebration of her husband.

But with the portrait of Lennon, the reception wasn't quite so warm. At White Cube's request, Breitz had set out to secure permission from the copyright holders of *John Lennon/Plastic Ono Band* prior to the first installations of the work at nonprofit museums in Newcastle and Vienna. Breitz wrote Yoko Ono to secure that permission. After a couple of months, she received a response from one of Ms. Ono's lawyers. "We are not able to grant the use of Mr. Lennon's image for your project," the e-mail informed. But Breitz didn't want permission to use Lennon's image. She wanted permission to engage with twenty-five fans singing his music. When Breitz responded with that correction, the lawyer informed her that he had not in fact personally reviewed her proposal. He was simply relaying the fact that Ms. Ono was not willing to grant the rights requested. A major international curator who knew Yoko and was a supporter of Breitz's work intervened on Breitz's behalf, suggesting that, as he understood the situation, Breitz could in fact have paid for the relevant copyrights and gone ahead with the project, but that out of respect, she was seeking Ono's permission and understanding. Ms. Ono wanted to hear more, but she disagreed with the curator about her freedom to make a cover without permission. "Permission," Ono insisted, "was vital, legally."

The curator described the proposal again. Ono asked to see it in writing. After reviewing it, her lawyers informed Breitz that she could use *John Lennon/Plastic Ono Band* in her project, but:

> Please note, clearance for the use of the actual musical compositions must be secured from the relevant publishers.[2]

Relieved (however naively), Breitz then asked White Cube's lawyers to start the process of securing "clearance" from the copyright holders for the compositions. Three months later, the lawyers representing Sony (holder of the rights to ten of the eleven songs on the album) quoted a standard fee of approximately $45,000 for one month's exhibition. Sony knew this was too much but wanted to set a baseline for the negotiations that would follow. They requested that the artist let them know the largest sum that she could afford. They wanted to see the project's budget.

Time, however, was running short. The exhibit was scheduled to open in Newcastle in a matter of weeks. After being pressed, the lawyers agreed to permit the work to be shown at this nonprofit institution without an agreement. They did the same for a nonprofit venue in Vienna three months later, but mentioned that Ms. Ono's lawyers wanted a formal agreement before any further exhibitions could go ahead.

A year after the request was originally made, it had still not been resolved. At the time of this writing, more than two years after the initial response, and after literally hundreds of hours of the lawyers', the museum executives', and Breitz's time, the rights holders have still not come to a final agreement. No one seems to have noticed that the value of the time spent dickering over these rights far exceeded any possible licensing fee. Economics didn't matter. A principle was at stake. As Ms. Ono had put it, "permission was vital, legally" before the love of twenty-five fans for the work of John Lennon could be explored publicly by another artist.

Gregg Gillis is a twenty-five-year-old biomedical engineer from Pittsburgh. He is also one of the hottest new artists in an emerging genre of music called "mash-up" or "remix." Girl Talk is the name of his one-man (and one-machine) band. That band has now produced three CDs. The best known, *Night Ripper*, was named one of the year's best by *Rolling Stone* and *Pitchfork*. In March 2007, his local congressman, Democrat Michael Doyle, took to the floor of the House to praise this "local guy made good" and his new form of art.

"New" because Girl Talk is essentially a mix of many samples drawn from many other artists. *Night Ripper*, for example, remixes between 200 and 250 samples from 167 artists. "In one example," Doyle explained on the floor of the House, "[Girl Talk] blended Elton John, Notorious B.I.G., and Destiny's Child all in the span of 30 seconds." Doyle was proud of this hometown wonder. He invited his colleagues to "take a step back" to look at this new form of art. "Maybe mash-ups," Doyle speculated, "are a transformative new art that expands the consumer's experience and doesn't compete with what an artist has made available on iTunes or at the CD store."

Doyle's comments helped fuel a flurry of media attention to Girl Talk. That, in turn, helped fuel some real anxiety among Girl Talk's distributors. For the defining feature of this mash-up genre is that the samples are remixed without any permission from the original artists. And if you ask any lawyer representing any label in America, he or she would quickly Ono-ize: "Permission is vital, legally." Thus, as Gillis practices it, Girl Talk is a crime. Apple pulled *Night Ripper* from the iTunes Music Store. eMusic had done the same a few weeks before. Indeed, one CD factory had refused even to press the CD.

Gillis had begun with music at the age of fifteen. Listening to electronic experimental music on a local radio station, he "discovered this world of people that could press buttons and make noise on pedals and perform it live." "It kind of blew my mind," he told me. At the age of sixteen he "formed a noise band—noise meaning very avant-garde music" for the time.[3]

Over the years, "avant-garde" moved from analog to digital—aka computers. Girl Talk the band was born in late 2000 on a Toshiba originally purchased for college. Gillis loaded the machine with audio tracks and loops. Then, using a program called Audio-Mulch, he would order and remix the tracks to prepare for a performance. I've seen Girl Talk perform live; his shows are as brilliant as his recorded remixes.

It wasn't long into the life of Girl Talk, however, that the shadow of Law Talk began to grow. Gillis recognized that his form of creativity didn't yet have the blessing of the law. Yet he told me, "I was never that fearful....I guess I was a little naive, but at the same time, it was just the world I existed in where you see these things every day. [And you] know you're going to be selling such a small number of albums that no one will probably ever take notice of it." There were of course famous cases where people did "take notice." Negativland, a band we'll see more of later in this book, had had a famous run-in with U2 and Casey Kasem after it remixed a recording of Kasem introducing the band on *American Top 40*. Gillis knew about this run-in. But as he explained to me in a way that reminded me of the days when I too thought the law was simply justice written nicely,

> I feel the same exact way now that I felt then. I think, just morally, that the music wasn't really hurting anyone. And there's no way anyone was buying my CD instead of someone else's [that I had sampled]. And...it clearly wasn't affecting the market. This wasn't something like a bootlegging case. I felt like if someone really had a problem with this then we could stop doing it. But I didn't see why anyone should.

Why anyone "should" was a question I couldn't answer. That someone would was a prediction too obvious to make. The "problem" would be raised not directly, but indirectly; not by filing a lawsuit against Girl Talk, but by calling up iTunes or another distributor and asking questions that made the distributor stop its distribution, and thus forcing this artist, and this art form, into obscurity. The "problem" of Girl Talk would be solved by making sure that any success of Girl Talk was limited. Keep it in Pittsburgh, and dampen the demand wherever you can, and maybe the "problem" would go away.

Gillis agrees the problem is going away. But for a very different reason. For the thing that Gillis does well, Gillis explained to me, everyone will soon do. Everyone, at least, who is passionate about music. Or, at least, everyone passionate about music and under the age of thirty.

> We're living in this remix culture. This appropriation time where any grade-school kid has a copy of Photoshop and can download a picture of George Bush and manipulate his face how they want and send it to their friends. And that's just what they do. Well, more and more people have noticed a huge increase in the amount of people who just do remixes of songs. Every single Top 40 hit that comes on the radio, so many young kids are just grabbing it and doing a remix of it. The software is going to become more and more easy to use. It's going to become more like Photoshop when it's on every computer. Every single P. Diddy song that comes out, there's going to be ten-year-old kids doing remixes and then putting them on the Internet.

"But why is this good?" I asked Gillis.

> It's good because it is, in essence, just free culture. Ideas impact data, manipulated and treated and passed along. I think it's just great on a creative level that everyone is so involved with the music that they like.... You don't have to be a traditional musician. You get a lot of raw ideas and stuff from people outside of the box who haven't taken guitar lessons their whole life. I just think it's great for music.

And, Gillis believes, it is also great for the record industry as well: "From a financial perspective, this is how the music industry can thrive in the future...this interactivity with the albums. Treat it more like a game and less like a product."

Gillis's point in the end, however, was not about reasons. It was about a practice. Or about the practice of this generation. "People are going to be forced—lawyers and...older politicians—to face this reality: that everyone is making this music and that most music is derived from previous ideas. And that almost all pop music is made from other people's source material. And that it's not a bad thing. It doesn't mean you can't make original content."

All it means—today, at least—is that you can't make this content legally. "Permission is vital, legally," even if today it is impossible to obtain.

SilviaO is a successful Colombian artist. For a time she was a songwriter and recording star, making CDs to be sold in the normal channels of Colombian pop music. In the late 1990s, she suffered a tragic personal loss, and took some time away from performing. When she returned to creating music, a close friend and developer for Adobe convinced her to try something different.

I saw her describe the experience outside a beautiful museum near Bogotá, at the launch of Creative Commons Colombia. (We'll see more of Creative Commons later. Suffice it to say for now that the nonprofit provides free copyright licenses to enable artists to mark their creative work with the freedoms they want it to carry. These licenses are then translated, or "ported," into jurisdictions around the world. When that porting is complete, the country "launches," making the new localized licenses available.) About a hundred people, mainly artists and twentysomethings, were gathered in an amphitheater next to the museum. SilviaO spoke in Spanish. A translator sitting next to me carried her words into English.

She told a story of donating an a cappella track titled "Nada Nada" ("Nothing Nothing") to a site Creative Commons runs called ccMixter. ccMixter was intended as a kind of Friendster for music. People were asked to upload tracks. As those tracks got remixed, the new tracks would keep a reference to the old. So you could see, for example, that a certain track was made by remixing two other tracks. And you could see that four other people had remixed that track.

SilviaO's track was a beautiful rendition of a song sung in Spanish, described on the ccMixter site as the story of "a girl not changing her ideas, dreams or way of life after engaging in a relationship." A few days after the track was uploaded, however, a famous mixter citizen, fourstones, remixed it—cutting up the Spanish into totally incomprehensible (but beautiful) gibberish, and retitling the mix "Treatment for Mutilation."

As she stood before those who had come to celebrate Creative Commons Colombia and described this "mutilation," I, the chairman of Creative Commons, began to sweat. I was certain she was about to attack remix creativity. A remixer had totally destroyed the meaning of her contribution. I was certain this was to become a condemnation of the freedom that I had thought we were all there to celebrate.

To my extraordinary surprise and obvious relief, however, SilviaO had no condemnation to share. She instead described how the experience had totally changed how she thought about creating music. Sure, the words were no longer meaningful. But the sound had taken on new meaning. As she told me later, "the song became more jazzy, and it opened the gate to understanding that maybe it was going to be more to treat my voice as an instrument and something completely independent from lyrics than I was used to before."[4]

Inspired by that remix, she wrote another track to be layered onto the first. Since then, she has added song after song to the ccMixter collection. Unlike Breitz's work or Girl Talk, all these remixes were legal. If "permission is vital, legally," then with this work, permission had already been given. The Creative Commons licenses had shifted the copyright baseline through the voluntary acts of copyright holders.

And for SilviaO, the act of creating had changed. Before, she sat in a studio, crafting work that would be broadcast, one to many. Now she was in a conversation with other artists, providing content they would add to, and adding content back. "I'm more talking with the musicians right now," she told me, "because I'm releasing my work and I know for sure, for many of them, they don't understand not even the words I am saying. [But] my voice is just another instrument, so all the options that they are playing with are completely their own. So there is more freedom....My voice," she explained, "was just a little bit—it was just a little part of the huge process that is happening now with this kind of creation. I was a little bit more free, because I didn't know how they were reacting.

"I became," she whispered, "a little bit more courageous."

If I asked you to shut your eyes and think about "the copyright wars," your mind would not likely run to artists or creators like these.

Peer-to-peer file sharing is the enemy in the "copyright wars." Kids "stealing" stuff with a computer is the target. The war is not about new forms of creativity, not about artists making new art. Congress has not been pushed to criminalize Girl Talk.

But every war has its collateral damage. These creators are just one type of collateral damage from this war. The extreme of regulation that copyright law has become makes it difficult, and sometimes impossible, for a wide range of creativity that any free society—if it thought about it for just a second—would allow to exist, legally. In a state of war, however, we can't be lax. We can't forgive infractions that might at a different time not even be noticed. Think "eighty-year-old grandma being manhandled by TSA agents," and you're in the frame for this war as well.

Collateral damage is the focus of this book. I want to put a spotlight on the stuff no one wants to kill—the most interesting, the very best of what these new technologies make possible. If the war simply ended tomorrow, what forms of creativity could we expect? What good could we realize, and encourage, and learn from?

I then want to spotlight the damage we're not thinking enough about—the harm to a generation from rendering criminal what comes naturally to them. What does it do to them? What do they then do to us?

I answer these questions by drawing a map of the change in what we could call cultures of creativity. That map begins at the turn of the last century. It is painted with fears from then about what our culture was becoming. Most of those fears proved correct. But they help us understand why much of what we seem to fear today is nothing to fear at all. We're seeing a return of something we were before. We should celebrate that return, and the prosperity it promises. We should use it as a reason to reform the rules that render criminal most of what your kids do with their computers. Most of all, we should learn something from it—about us, and about the nature of creativity.

PART ONE

CULTURES

ONE

CULTURES OF OUR PAST

O n a humid day in June 1906, one of America's favorite composers climbed the steps of the Library of Congress to testify
about the status of copyright law in America. John Philip Sousa
was a critic of the then relatively lax United States copyright system.
He had come to Washington to ask that Congress "remedy a serious defect in the . . . law, which permits manufacturers and sellers of
phonograph records . . . to appropriate for their own profit the best
compositions of the American composer without paying a single
cent therefor"—a form of "piracy" as he called it.[1]

Sousa's outrage is not hard to understand. Though he was a
famous conductor, some of Sousa's income came from the copyrights he had secured in the work he had composed and arranged.
Those copyrights gave him an exclusive right to control the public performance of his work; any reproduction of sheet music to
support that public performance; and any arrangements, or other
work, "derived" from his original work. This mix of protections
was crafted by Congress to reward artists for their creativity by creating incentives for artists to produce great new work.

The turn of the century, however, brought an explosion of

technologies for creating and distributing music that didn't fit well within this old model of protection. With these new technologies, and for the first time in history, a musical composition could be turned into a form that a machine could play—the player piano, for example, or a phonograph. Once encoded, copies of this new musical work could be duplicated at a very low cost. A new industry of "mechanical music" thus began to spread across the country. For the first time in human history, with a player piano or a phonograph, ordinary citizens could access a wide range of music on demand. This was a power only kings had had before. Now everyone with an Edison or an Aeolian was a king.

The problem for composers, however, was that they didn't share in the wealth from this new form of access. Mechanical music may have in one sense "copied" their work. But as most courts interpreted the Copyright Act, whatever "copy" these machines made was not the sort of copy regulated by the law. This angered many composers. Some, such as Sousa, resolved to do something about it. His trip to Capitol Hill was just one part of his extensive (and ultimately successful) campaign.

My interest in Sousa's testimony, however, has little to do with his (to us, today) obviously sensible plea. It is instead a point that may have been obvious to him, then, but that has largely been forgotten by us, now. For as well as complaining about the "piracy" of mechanical music, Sousa also complained about the cultural emptiness that mechanical music would create. As he testified:

> When I was a boy...in front of every house in the summer evenings you would find young people together singing the songs of the day or the old songs. Today you hear these infernal machines

going night and day. We will not have a vocal cord left. The vocal cords will be eliminated by a process of evolution, as was the tail of man when he came from the ape.[2]

"We will not have a vocal cord left."

John Philip Sousa was obviously not offering a prediction about the evolution of the human voice box. He was describing how a technology—"these infernal machines"—would change our relationship to culture. These "machines," Sousa feared, would lead us away from what elsewhere he praised as "amateur" culture. We would become just consumers of culture, not also producers. We would become practiced in selecting what we wanted to hear, but not practiced in producing stuff for others to hear.

So why would one of America's most prominent professional musicians criticize the loss of amateur music?

Sousa's fear was not that the quality of music would decline as less was produced by amateurs and more by professionals. Instead, his fear was that culture would become less *democratic:* not in the sense that people would vote about what is, or is not, good culture, but in a sense that MIT professor Eric von Hippel means when he argues that innovation today is becoming more "democratized."[3] In the world Sousa feared, fewer and fewer would have the access to instruments, or the capacity, to create or add to the culture around them; more and more would simply consume what had been created elsewhere. Culture would become the product of an elite, even if this elite, this cultural monarchy, was still beloved by the people.

Indeed, he believed this change was already happening. As he recounted:

Last summer…I was in one of the biggest yacht harbors of the world, and I did not hear a voice the whole summer. Every yacht had a gramophone, a phonograph, an Aeolian, or something of the kind. They were playing Sousa marches, and that was all right, as to the artistic side of it, but they were not paying for them, and, furthermore, they were not helping the technical development of music.[4]

This decline in participation, Sousa argued, would translate into a decline in the spread of tools to create music:

This wide love for the art springs from the singing school, secular or sacred; from the village band, and from the study of those instruments that are nearest the people. There are more pianos, violins, guitars, mandolins, and banjos among the working classes of America than in all the rest of the world, and the presence of these instruments in the homes has given employment to enormous numbers of teachers who have patiently taught the children and inculcated a love for music throughout the various communities.[5]

"And what is the result" of this loss of "amateurs"? Sousa asked.

The child becomes indifferent to practice, for when music can be heard in the homes without the labor of study and close application, and without the slow process of acquiring a technique, it will be simply a question of time when the amateur disappears entirely….*[T]he tide of amateurism cannot but recede*, until there will be left only the mechanical device and the professional executant.[6]

"The tide of amateurism cannot but recede"—a bad thing, this professional believed, for music and for culture.

Sousa was romanticizing culture in a way that might remind the student of American history of Thomas Jefferson. Jefferson romanticized the yeoman farmer.[7] He would be sickened by the modern corporate farm that has displaced his yeoman hero. But his repulsion would have little to do with the efficiency of food production, or even the quality of the food produced. Instead, he would object to the effect of this change on our democracy. Jefferson believed that the ethic of a yeoman farmer—one practiced in the discipline of creating according to an economy of discipline, as any farmer on the edge of civilization in eighteenth-century America would—was critical to democratic self-governance. Yeoman self-sufficiency was thus not a virtue because it was an efficient way to make food. Yeoman self-sufficiency was a virtue because of what it did to the self, and in turn, what it did to democratic society, the union of many individual selves.

Sousa's take on culture was similar. His fear was not that culture, or the actual quality of the music produced in a culture, would be less. His fear was that people would be less connected to, and hence practiced in, creating that culture. Amateurism, to this professional, was a virtue—not because it produced great music, but because it produced a musical culture: a love for, and an appreciation of, the music he re-created, a respect for the music he played, and hence a connection to a democratic culture. If you want to respect Yo-Yo Ma, try playing a cello. If you want to understand how great great music is, try performing it with a collection of amateurs.

RW Culture Versus RO Culture

In the language of today's computer geeks, we could call the culture that Sousa celebrated a "Read/Write" ("RW") culture:* in Sousa's world (a world he'd insist included all of humanity from the beginning of human civilization), ordinary citizens "read" their culture by listening to it or by reading representations of it (e.g., musical scores). This reading, however, is not enough. Instead, they (or at least the "young people of the day") add to the culture they read by creating and re-creating the culture around them. They do this re-creating using the same tools the professional uses—the "pianos, violins, guitars, mandolins, and banjos"—as well as tools given to them by nature—"vocal cords." Culture in this world is flat; it is shared person to person.[8] As MIT professor Henry Jenkins puts it in his extraordinary book, *Convergence Culture,* "[T]he story of American arts in the 19th century might be told in terms of the mixing, matching, and merging of folk traditions taken from various indigenous and immigrant populations."[9]

Sousa's fear was that this RW culture would disappear, be displaced by—to continue the geek-speak metaphor—an increasingly "Read/Only" ("RO") culture: a culture less practiced in performance, or amateur creativity, and more comfortable (think: couch) with simple consumption. The fear was not absolute: no one feared that all nonprofessional creativity would disappear. But certainly its

---

* The analogy is to the permissions that might attach to a particular file on a computer. If the user has "RW" permissions, then he is allowed to both read the file and make changes to it. If he has "Read/Only" permissions, he is allowed only to read the file.

significance and place within ordinary society would change. RW creativity would become less significant; RO culture, more.

As one reflects upon the history of culture in the twentieth century, at least within what we call the "developed world," it's hard not to conclude that Sousa was right. Never before in the history of human culture had the production of culture been as professionalized. Never before had its production become as concentrated. Never before had the "vocal cords" of ordinary citizens been as effectively displaced, and displaced, as Sousa feared, by these "infernal machines." The twentieth century was the first time in the history of human culture when popular culture had become professionalized, and when the people were taught to defer to the professional.

The "machines" that made this change possible worked their magic through tokens of RO culture—recordings, or performances captured in some tangible form, and then duplicated and sold by an increasingly concentrated "recording" industry. At first, these tokens were physical—player-piano rolls, then quickly phonographs. In 1903, "the Aeolian Company had more than 9,000 [player-piano] roll titles in their catalog, adding 200 titles per month."[10] During the 1910s, "perhaps 5% of players sold were reproducing pianos." At one point in the 1920s, a majority of the pianos made in America had a player unit included.[11]

Phonographs shared a similar growth. In 1899, 151,000 phonographs were produced in the United States.[12] Fifteen years later, that number had more than tripled (to approximately 500,000 units). Record sales in 1914 were more than 27 million.[13] But for most of the 1920s, sales stayed above 100 million copies.[14] By the late 1920s, between 33 percent and 50 percent of all households had a record player.[15] Nineteen twenty-nine was the peak for record sales in the United States[16] before the Depression burst this and many other cultural bubbles.

But as radio technology improved, physical tokens of RO culture faced competition from tokens that were more virtual—what we call "broadcasts." To compete, phonograph manufacturers cut prices. "In 1925, Victor dropped the price of its $1.50 single-side Red Seal records to 90 cents, and cut its $1.00 records to 65 cents."[17] But as Philip Meza describes, "the price cuts did not work, and sales continued to fall.... In 1919, 2.2 million phonographs were sold. In 1922, fewer than 600,000...."[18]

Competition drove the producers of physical tokens to produce higher-quality tokens. That in turn drove the demand for higher-quality radio—a demand that inspired Edwin Howard Armstrong to invent, the FCC to allow, and RCA to deploy FM radio.[19] Radio, however, soon faced its own competition from a new form of broadcast—television. The cycle then continued.

The twentieth century was thus a time of a happy competition among RO technologies. Each cycle produced a better technology; each better technology was soon bested by something else. The record faced competition from tapes and CDs; the radio, from television and VCRs; VCRs, from DVDs and the Internet.

By the turn of the twenty-first century, this competition had produced extraordinary access to a wide range of culture. Never before had so much been available to so many. It also produced an enormously valuable industry for the American economy and others. In 2002, the publishing industry alone (excepting the Internet) had revenues close to $250 billion.[20] In the same year, the revenue for broadcasting (again excepting the Internet) was almost $75 billion.[21] The revenue to the motion-picture and sound-recording industries was close to $80 billion.[22] And according to the Motion Picture Association of America,

> Core Copyright industries are responsible for an estimated 6% of the nation's total GDP totaling $626 billion a year. Copyright industries had an annual employment growth rate of 3.19% per year—a rate more than double the annual employment growth rate achieved by the economy as a whole.[23]

RO culture had thus brought jobs to millions. It had built superstars who spoke powerfully to millions. And it had come to define what most of us understood culture, or at least "popular culture," to be.

Limits in Regulation

Before RO culture carries us away, however, return for a moment to Sousa. For there was a second aspect to the culture that Sousa described that we should also notice here. This was the relationship between culture and the particular form through which we regulate culture—copyright law. It was about the limits on that regulation.

For his time, Sousa was a copyright extremist. He had come to Washington to push for (what was perceived by many to be) a radical increase in the reach of copyright. The push was opposed by many in the business world and many antiregulation idealists.

Yet Sousa's extremism still knew an important limit, a place where copyright law would reach too far. That limit got revealed

midway through his testimony. As he testified Sousa was interrupted by Congressman Frank Dunklee Currier, a Republican
from New Hampshire. After Sousa described the "young people
together singing the songs of the day and the old songs," Currier
asked:

> Currier: Since the time you speak of, when they used to be
>   singing in the streets ... the law has been [changed] ... to
>   prohibit that. Is not that so?
> Sousa: No, sir; you could always do it.
> Currier: Any public performance is prohibited, is it not, by
>   that law?
> Sousa: You would not call that a public performance.
> Currier: But any public performance is prohibited by the law
>   of 1897?
> Sousa: Not that I know of at all. I have never known that it
>   was unlawful to get together and sing.[24]

Though the record doesn't indicate it, one imagines laughter followed Sousa's comment. And anyway, Currier was not being serious. He was not a copyright extremist. Indeed, quite the opposite.
Currier was an "intellectual property" skeptic, unconvinced of the
need for this government-backed monopoly to interfere with inven-
tions or the arts. The aim of his question was to embarrass Sousa
for Sousa's (from Currier's view) extremism.[25] He wanted to suggest
the law had already gone too far and didn't need to go any further.

The effort backfired. Sousa didn't believe that every use of culture should be regulated. Indeed, he thought it ridiculous to imagine a world where it was "unlawful to get together and sing." That
part of culture (a critical part if amateur culture was to survive)

must be left unregulated, Sousa believed, even if another part of culture (the part where commercial entities profited from creative works) needed to be regulated more. Even for this extremist, copyright law had a limit.

Keep these two ideas in mind as we turn to the argument that follows: one, the importance of "amateur" creativity, producing an RW culture; two, the importance of limits in the reach of copyright's regulation, leaving free from regulation this amateur creativity.

In the balance of this book, my hope is to revive these two Sousarian sensibilities. As we look back at our history, the dominance of the radically different culture (and the culture of regulating culture) of the last forty years is likely to obscure the view of a much longer tradition that lived before it. That much longer tradition has value for us today. For the conditions that made its best part possible are now returning. And ironically for Mr. Sousa, they are returning precisely because of a new generation of (as professional musicians today call them) "infernal machines." These new infernal machines, however, will enable an RW culture again. And if permitted by the industries that now dominate the production of culture (and that exercise enormous control over Congress, which regulates that culture), they could also encourage an enormous growth in economic opportunity for both the professional and the amateur, and for all those who benefit from both forms of creativity.

TWO

CULTURES OF OUR FUTURE

The "copyright wars" have lead many to believe that the choice we face is all or nothing. Either Hollywood will win or "the Net" will win. Either we're about to lose something important that we've been, or we're going to kill something valuable that we could be. Whoever wins, the other must lose.

This simple framing creates a profound confusion. For there need be no trade-off between the past and the future. Instead, all the evidence promises an extraordinary synthesis of the past and the present to create a phenomenally more prosperous future. This future need not be either less RO or more RW: it could be both. And much more interesting (to those focused on the economy, at least), this future could see the emergence of a form of economic enterprise that has been relatively rare in our past, but that promises extraordinary economic opportunity: what I call the "hybrid."

In the chapters that follow, I want to map this future. I start with what simply continues the twentieth century—a story of how the Internet extends RO culture beyond the unavoidable limits of twentieth-century technology. I then show just how the same technologies that encourage RO culture could also encourage the revival of the RW creativity that Sousa celebrated. Finally, I describe the most interesting change that I believe we're going to see—the "hybrid"—that will increasingly define the industries of culture and innovation. All three changes, if allowed, will be valuable and important. All three should be encouraged.

THREE

RO, EXTENDED

There's a part of culture that we simply consume. We listen to music. We watch a movie. We read a book. With each, we're not expected to do much more than simply consume.* We might hum along with the music. We might reenact a dance from a movie. Or we might quote a passage from the book in a letter to a friend. But in the main, this kind of culture is experienced through the act of consumption. There's a beginning, a middle, and an end to that consumption. Once we've finished it, we put the work away.

This is the stuff at the core of RO culture. And while of course the stuff was not born with the "infernal machines" that Sousa lamented (in our tradition it was Gutenberg who gave birth to the most significant spread of tokens of RO culture), my focus for the moment will be on the RO culture that Sousa did lament: the tokens of RO culture that get processed and performed by machines, capturing and spreading music, and the spoken word, and eventually, images and film.

---

* Of course, as Candice Breitz and many others argue, there's nothing "simple" in consuming, but put those complications aside for the moment.

For most of the twentieth century, these tokens were analog. They all therefore shared certain limitations: first, any (consumer-generated) copy was inferior to the original; and second, the technologies to enable a consumer to copy an RO token were extremely rare. No doubt there were recording studios aplenty in Nashville and Motown. But for the ordinary consumer, RO tokens were to be played, not manipulated. And while they might legally be shared, every lending meant at least a temporary loss for the lender. If you borrowed my LPs, I didn't have them. If you used my record player to play Bach, I couldn't listen to Mozart.

These are the inherent—we could say "natural"—limitations of analog technology. From the consumer's perspective, they were bugs. No consumer ever bought a record player because he couldn't copy the records.

But from the perspective of the content industry, these limitations in analog technology were not bugs. They were features. They were aspects of the technology that made the content industry possible. For this nature limited the opportunity for consumers to compete with producers (by "sharing"). And its imperfections drove demand for each new generation of technology. Record companies thus sold bits of culture, embedded in vinyl records, then in eight-track tapes, then in cassette tapes, and then in CDs. With each new format, there was a wave of new demand (often for the very same work). The same with film. Film companies distributed films to theaters, and then films to videocassettes, and then films to DVDs. The business model of both these distributors of RO culture depended upon controlling the distribution of copies of culture. The nature of analog tokens of RO culture supported this business model by making it very difficult to do much differently.

The law supported this business model. The law, for example, forbade a consumer from making ten thousand copies of his favorite LP to share with his friends.[1] But it wasn't really the law that mattered most in stopping this form of "piracy." It was the economics of making a copy in the world of analog technology. At least among consumers, it was this nature of the LP that really limited the consumer's ability to be anything other than "a consumer."

Nature Remade

Digital technology changed this "nature." With the introduction of digital tokens of RO culture and, more important, with the widespread availability of technologies that could manipulate digital tokens of RO culture, digital technology removed the constraints that had bound culture to particular analog tokens of RO culture. As I've described in a different context,[2] we could say that while the *code* of an LP record protected it from duplication, the *code* of a digital copy of that record does not. The *code* of an analog videocassette effectively limited the number of times it could be played (before the tape wore out, for example). The *code* of a digital copy of that film does not. The "natural" constraints of the analog world were abolished by the birth of digital technology. What before was both impossible and illegal is now just illegal.

When the content industry recognized this change, it was terrified. Digital tokens of RO culture would no longer conspire with the content industry to protect that industry's business model. Unlike analog technologies and analog tokens of RO culture, digital technologies would instead conspire with the enemy—at least, the enemy of this particular business model. By the mid-1990s, the industry came to fully recognize this enemy. By the late 1990s, it had hatched a strategy to fight it.

And thus were born the copyright wars. In September 1995, the content industry, working with the U.S. Department of Commerce, began to map a strategy for protecting a business model from digital technologies.[3] In 1997 and 1998, that strategy was implemented in a series of new laws designed to extend the life of copyrighted work,[4] strengthen the criminal penalties for copyright infringement,[5] and punish the use of technologies that tried to circumvent digital locks placed on digital content.[6]

This legislation was soon complemented by aggressive litigation. First the lawyers targeted commercial entities like MP3.com and Napster.[7] Then they targeted ordinary citizens, charging them with downloading music or enabling others to do the same.[8] The federal system was flooded with claims based upon federal copyright law. According to one site that monitors lawsuits filed by the Recording Industry Association of America, as of June 2006, the RIAA had sued 17,587 people, including a twelve-year-old girl and a dead grandmother.[9] A year later, the RIAA had sent around 2,500 prelitigation letters to twenty-three more universities across the nation, threatening action based upon students' allegedly illegal downloading of copyrighted content.[10] These aggressive legal threats have coincided with a 250 percent increase in copyright litigation in the federal courts in six years.[11] A similar pattern has spread overseas. The International Federation of the Phonographic Industry (European cousin to the RIAA) reported suing more than ten thousand people in eighteen countries by the end of 2006. It promised many more suits in 2007.[12]

By the turn of the century, the industry's view had become simple and dire: As never before (at least since the last time),[13] the content industry was threatened by new technologies. And unless the government launched a massive effort to regulate the use and spread of these technologies, the rise of digital technologies would mean the fall of much of the content industry.

The numbers then were at least consistent with the content industry's argument: By the first half of 2002, world sales of recorded music had fallen by 9.2 percent in dollar value, and unit shipments were down 11.2 percent. Worldwide, the recording industry suffered its third straight year of declining sales. Sony told investors it expected music revenues to fall an additional 13–15 percent in 2003: "In the United States, sales had also declined steadily over the previous three years, with sales of recorded music falling 8.2 percent in dollar value and 11.2 percent in unit shipments."[14] The labels blamed "piracy" for "an estimated $5 billion loss in 2002" alone.[15] More recent statistics are, if anything, worse.[16]

Most in the industry—at least circa 2002—believed that "piracy" was unavoidable given the "nature" of digital technologies. Most thus believed the industry faced a choice: drive digital to the periphery and save the industry, or allow it to become mainstream, and watch the industry fail.

Re-remaking Nature

Then Steve Jobs taught them differently. For at the height of the frenzy of this war against "piracy," Jobs demonstrated in practice what many had been arguing in theory: that the only *nature* of digital technology is that it conforms to how it is coded. The technologies of the Internet were originally coded in a way that enabled free, and perfect, copies, that nature could be changed by a different code, with different permissions built in. Thus, digital tokens of RO culture could be recoded with at least enough control to restore a market in their distribution. That market could, Jobs demonstrated, compete effectively with the "free" distribution of the Internet.

The iTunes Music Store was the proof. Launched in 2003, more than 1 billion songs were downloaded within three years, 2.5 billion within four.[17] And while iTunes music was digital, iTunes tokens of digital culture contained a technology to limit their (re)distribution. Code (called FairPlay, a kind of Digital Rights Management, or DRM technology) was used to remake the code of digital tokens of RO culture. This remade code was enough to get a reluctant content industry to play along.

Apple's iTunes wasn't the first to embed DRM in content.[18] It was just the smartest. Jobs understood that the record companies would demand some control. The success of iTunes (and more important, of the iPod conveniently tied to it) came from the fact that "some control" could be less than "perfect control." You couldn't *easily* spread iTunes content to everyone on the Web—though if you hunted around a bit on the Net, you'd find all the code you could want to liberate iTunes. DRM was just a speed bump: it slowed illegal use just enough to get the labels to buy in.

I'm not saying it was Jobs's genius alone that brought the content industry around. An important legal lever was being deployed at the same time in the Napster case. Recall that the record companies had sued Napster because of the "piracy" it enabled. Napster had countersued the record labels, charging that they had an agreement among themselves not to sell content to the digital platform.[19] The labels needed cover from this charge, and an experiment with an operating system holding no more than 5 percent of the market seemed safe enough. Thus was iTunes born.

But whatever the motivation, or the mix of motivations, iTunes' success supported the idea that a wide range of content might be sold digitally on the same model that defined the content industry of the twentieth century: by metering the number of copies sold. iTunes quickly expanded its offerings to books, then music videos and TV shows, and, finally, movies. Others followed a similar path—offering different models for selling culture, but all still *selling* culture nonetheless. eMusic convinced independent labels to sell downloads without any DRM. Rhapsody sold DRM'd downloads in a subscription model. The key with each successful example was to find a balance between access and control that would satisfy both the consumers and the creators. This mix of models soon convinced a skeptical industry that RO culture had a twenty-first-century future. And soon into the century, there was a revival of investment to find ways to better spread and exploit an RO market in a digital age.

The potential is not hard to envision; the businesses are just beginning to emerge now. If the twentieth century made culture generally accessible, the twenty-first will make it universally accessible. As the cost of inventory drops, the mix of inventory increases—the lesson of the Long Tail, which we'll consider more in chapter 6. As the mix increases, the diversity of culture that can flourish in the digital age grows. Think of all the books in the Library of Congress. Now imagine the same diversity of music, video, and images. And then imagine all of it accessible, in an instant, by anyone, anywhere. No doubt there are lots of hurdles to overcome to get to this world. But the hurdles are not technical. As we'll see in chapter 9, they are just regulatory. And if these regulatory burdens can be reduced, a new industry of RO culture can flourish. A hundred years from now, if it is allowed to flourish, we will see its relationship to the twentieth century as we see the relationship between the Boeing 777 and the work of the Wright brothers or Alberto Santos-Dumont. This is the extraordinary potential for RO culture in a digital age.

Recoding Us

As these businesses grow, they change not only business. They also change us. They change how we think about access to culture. They change what we take for granted.

For example: during the twentieth century, our access to television and movies was different from our access to books. With television and movies, the viewer had to conform his schedule to the schedule of the distributor. So much was required by the technology; so much came to seem natural. "Channels" were tools to *channel* people into watching one mix of content rather than another. A smart scheduler tried to keep an audience by varying the mix so as to prevent the "viewer" from wandering to another channel.

During the same period, however, books were accessed differently. With books, the "natural" expectation (in the twentieth century at least) was that the content was accessible on our schedule. When we walked into a library, we expected to get what we wanted, then. If the library didn't have it, we expected it to get what we wanted relatively quickly through interlibrary loan. If a librarian had told you as you entered the library, "I'm sorry, in the afternoon we offer only nonfiction. If you'd like to read some fiction, come back after five p.m.," you would have been incensed. The idea that the library gets to say when and what I read is outrageous. Or put differently, it would have been considered outrageous for any library or bookstore or publisher to exercise the same control over access to books that television stations and film distributors exercised over film and video.

In the twenty-first century, television and movies will be book-i-fied. Or again, our expectations about how we should be able to access video content will be the same as the expectations we have today about access to books. The idea that you would conform your schedule to a distributor's will seem increasingly ridiculous. The idea that you would have to wait till "prime time" to watch prime television will seem just fascist. Freedom will mean freedom to choose to watch what you want when you want, just as freedom to read means the freedom to read what you want when you want. In both cases, not necessarily *for free*. But in both cases, according to your schedule, not the schedule of someone else.

We can see this most clearly in our kids, who think it "just dumb" that an episode of a favorite TV series is not available whenever they want to see it. And even older sorts begin to understand this sense, as the DVRs like ReplayTV and TiVo become increasingly common. More and more, even to old folks like me, it seems astonishing to remember a time when to watch a television show, you had to synchronize your schedule to the schedule of the broadcaster. Absurd that if you missed an episode, that was it. There was no chance—at least that season—for a repeat.

The expectation of access on demand builds slowly, and it builds differently across generations. But at a certain point, perfect access (meaning the ability to get whatever you want whenever you want it) will seem obvious. And when it seems obvious, anything that resists that expectation will seem ridiculous. Ridiculous, in turn, makes many of us willing to break the rules that restrict access. Even the good become pirates in a world where the rules seem absurd.

I saw this dynamic in myself with the 2007 Academy Awards. For weird and accidental reasons (meaning, I don't hang out with movie stars), I had two friends nominated for an Oscar in 2007. I was thus desperate to watch the awards. But that year, I was on sabbatical in Germany, and not desperate enough to get up at 3 a.m. to watch hours of Hollywood self-promotion. So I programmed a VCR to record the show, and went to bed expecting to awaken and watch the results.

I'm not a technical genius, but I'm also not an idiot. Nonetheless, as seems always to be the case, the VCR didn't record. So though I could read that both of my friends had indeed won Oscars, I was extremely disappointed that I couldn't watch them win.

My first reaction was to turn to the Web site of the Academy Awards. The site had fancy advertisements that changed with every click you made, and tons of content. They must, I thought, have video of the awards ceremony available to be streamed. It's 2007, I thought. And the Academy Awards ceremony is a wasting asset: while many will care about the program in February 2007, almost no one will care in March.

But the site didn't have the actual ceremony available for free (or "free," since all content on the site was run with ads surrounding it) or even to purchase. So I turned to iTunes, willing to pay whatever it would charge to download the awards ceremony. But again, no luck. iTunes didn't have it. I then extended my search to a number of other obvious places where the program might be for sale. Yet again, no luck.

So then I did something I just don't do—I went to YouTube to see who might have at least clips that might show my friends accepting their awards. Within five minutes, I had found clips with both friends, which I watched with utter joy.

Many people did the same as I (though I take it not for the same reason). And many took those clips and blogged them—adding commentary, or criticism, or praise for the works celebrated at the awards. I did too, adding links to the YouTube clips on my blog in an entry the next day bragging about my Oscar-winning friends. But then I read about legal action being initiated against bloggers and YouTube users who had distributed parts of the awards. And while, as a lawyer, I understood precisely the claim the content owners had, as a citizen of the twenty-first century, I was still astonished. Though this instinct can't be justified as a matter of (at least today's) law, it is the essence of practical reason in the digital age: if you don't want your stuff stolen, make it easily available. YouTube is a picture of unmet demand. And indeed, when I've tried to find clips of important breaking news on YouTube, the only times I've failed have been when the content provider has made the same content available on its own site. Access is the mantra of the YouTube generation. Not necessarily free access. Access.

Digital technologies will thus shift the expectations surrounding access. Those changes will change other markets as well. Think of the iPod—perfectly integrating all forms of RO culture into a single device. That integration will increasingly lead us to see the device not as music player, or video player, but as a universal access point, facilitating simple access to whatever we want whenever we want. Many devices will compete to become this device. And that competition is certain to produce an extraordinarily efficient tool to facilitate, and meter, and police our access to a wide range of culture.

This change, in turn, will change other markets as well. Think about a hotel room: at high-quality hotels, there is now fierce competition to provide extremely high-quality televisions. *Why* is beyond me. What chance is there that in the thirty minutes I have before I go to sleep I will find something just starting on the 150 channels the hotel provides that I actually want to watch? From my perspective, at least, this $2,000 flat-screen television is a useless suck of space in a hotel room.

But as the universal access devices I've described get perfected, the same competition that drives hotels to spend thousands to give me beautiful access to the shopping channel will drive them to provide a simple way to connect my access device to their projector. Count on a future of simple docking devices that amplify or project content accessed through an iPod-like device. Hotels (and restaurants, airplanes, and bars) will then focus on supplying great infrastructure. The iUser brings the content.

Users will thus demand access at any time, to everything (think: Library of Congress). And technologies will develop to provide or meter or police that access (think: the iPod, 2020). But then which of these three models for access will it be? Will these devices simply provide access, either by simply holding the content, or by enabling the user to tune into a particular channel? Or like a jukebox, will they meter access, deducting a fee for every download or play? Or like a soldier at a military base, will they monitor the content being accessed, and block access without the proper credentials?

The easy, and to some degree true, answer is that they will do all three. But the interesting part is how significant the first of those three will be, and how insignificant the third. My sense is that digital technology will enable market support for a much wider range of "free" content than anyone expects now (where "free" simply means without charge); and digital technologies will continue to resist models that depend upon the heavy policing by its owners to protect against "unauthorized use." The quick disappearance of

DRM for music is evidence of the latter point.[20] I, however, want to focus here on the former.

The model for commercial broadcasting in the twentieth century was ad-supported "free" content. The limitations of the technology of the twentieth century restricted the ways in which ads might support free content. Programs were interrupted. Ads that roughly matched the demographic of the program's audience were broadcast. In a world of relatively few channels, those ads had sufficient penetration to make them pay (both the networks and the advertiser).

The limits in that technology are obvious: The advertiser has to broadcast to a wide range of people; the ability to target ads is relatively weak. The advertiser can't really know who saw the ad or what they did when they saw it. And the advertiser is constantly aware that his message is viewed as an intrusion. When ads came every thirty minutes or so, for many, they were a welcome break. But when 25 percent of broadcasting time is advertisement, they are a perpetual annoyance. (Indeed, at this frequency, you can begin to understand why there's a market to buy "free" TV: if eighteen minutes of every hour is advertisements, then even if you value your time at the minimum wage, it would pay to spend $1.99 to avoid watching the commercials.)

But just as the limitations of analog RO culture were eliminated by digital technologies, so too the limitations of twentieth-century advertising can be eliminated by twenty-first-century digital technology. And as the lessons of this change get spread, content providers will increasingly recognize that free access pays. Free access is a means to gather extremely valuable data about the viewer. That data can translate into much more effective advertising techniques.

The point is obvious when you think about Amazon. Amazon

knows me intimately because it watches me more carefully than does any thing or person in the world. No one could pick a better list of things I'm likely to want to buy. That's because Amazon sees what I buy. It learns from the patterns of other customers what the sort who buys as I do is likely to want to buy next. It has built a thick profile of my preferences. And I'm very happy to listen to it when it suggests something I might be interested in.

So imagine a network with the same data about you. (I know, privacy alarms are going off, and that's an important issue of course, but it's not the issue for this book.)[21] Imagine that by watching all the YouTube clips you browsed through or the shows you actually paused to watch, the network began to build a profile of your preferences as rich as Amazon's. And as it developed this profile, this network could now market more effectively than any network today. Access is what produces this value. Limiting access limits it.

This third point will be recognized soon, and not because dweeby professors write about it. That's the great thing about markets: there's never a need to lecture a competitive market. Markets are driven to find value through competition with others. No doubt, every age will be marked with battles waged by the previous generation's giants. But the giants always fall to a better way of making money. And the RO culture that digital technologies will support will provide lots of new ways for content producers to make money. "As if by an invisible hand," this market will radically change the nature of access to culture in the next ten years. As a result, our children will be unable to understand a world where Thursday at 10 p.m. was more significant in cultural terms than Friday at 5 a.m.

By invoking Adam Smith's "invisible hand," I don't mean to say that policy makers have nothing to worry about here. Smith's

Wealth of Nations teaches us about the phenomenal power of markets to adjust. But these markets adjust, as Yochai Benkler's *The Wealth of Networks* powerfully teaches, in light of the baseline allocation of rights. Policy makers must assure that rights are not allocated in a way that distorts or weakens competition. A costly overlay of spectrum rights, for example, or an inefficient market of copyrights, can stifle competition and drive markets to unnecessary concentration. These factors must be regulated by policy makers. They will not be "solved" by an invisible hand.

But for my purposes here, the most important policy mistake is one that stifles the Sousarian instinct: a policy driven by the view that the only way to protect RO culture is to render RW culture illegal. That choice is a false choice. In the next chapter, I want to sketch a future for RW culture that might motivate us to see just why we should avoid this false choice.

FOUR

RW, REVIVED

One of my closest (if most complicated) friends at college was an English major. He was also a brilliant writer. Indeed, in every class in which writing was the measure, he did as well as one possibly could. In every other class, he, well, didn't.

Ben's writing had a certain style. Were it music, we'd call it sampling. Were it painting, it would be called collage. Were it digital, we'd call it remix. Every paragraph was constructed through quotes. The essay might be about Hemingway or Proust. But he built the argument by clipping quotes from the authors he was discussing. Their words made his argument.

And he was rewarded for it. Indeed, in the circles for which he was writing, the talent and care that his style evinced were a measure of his understanding. He succeeded not simply by stringing quotes together. He succeeded because the salience of the quotes, in context, made a point that his words alone would not. And his selection demonstrated knowledge beyond the message of the text. Only the most careful reader could construct from the text he read another text that explained it. Ben's writing showed he was an insanely careful reader. His intensely careful reading made him a beautiful writer.

Ben's style is rewarded not just in English seminars. It is the essence of good writing in the law. A great brief seems to say nothing on its own. Everything is drawn from cases that went before, presented as if the argument now presented is in fact nothing new. Here again, the words of others are used to make a point the others didn't directly make. Old cases are remixed. The remix is meant to do something new. (Appropriately enough, Ben is now a lawyer.)

In both instances, of course, citation is required. But the cite is always sufficient payment. And no one who writes for a living actually believes that any permission beyond that simple payment should ever be required. Had Ben written the estate of Ernest Hemingway to ask for permission to quote *For Whom the Bell Tolls* in his college essays, lawyers at the estate would have been annoyed more than anything else. What weirdo, they would have wondered, thinks you need permission to quote in an essay?

So here's the question I want you to focus on as we begin this chapter: Why is it "weird" to think that you need permission to quote? Why would (or should) we be "outraged" if the law required us to ask Al Gore for permission when we wanted to include a quote from his book *The Assault on Reason* in an essay? Why is an author annoyed (rather than honored) when a high school student calls to ask for permission to quote?

The answer, I suggest, has lots to do with the "nature" of writing. Writing, in the traditional sense of words placed on paper, is the ultimate form of democratic creativity, where, again, "democratic" doesn't mean people vote, but instead means that everyone within a society has access to the means to write. We teach everyone to write—in theory, if not in practice. We understand quoting is an essential part of that writing. It would be impossible to construct and support that practice if permission were required every time a quote was made. The freedom to quote, and to build upon, the words of others is taken for granted by everyone who writes. Or put differently, the freedom that Ben took for granted is perfectly natural in a world where everyone can write.

Writing Beyond Words

Words, obviously, are not the only form of expression that can be remixed in Ben's way. If we can quote text from Hemingway's *For Whom the Bell Tolls* in an essay, we can quote a section from Sam Wood's film of Hemingway's *For Whom the Bell Tolls* in a film. Or if we can quote lyrics from a Bob Dylan song in a piece about Vietnam, we can quote a recording of Bob Dylan singing those lyrics in a video about that war. The act is the same; only the source is different. And the measures of fairness could also be the same: Is it really just a quote? Is it properly attributed? And so on.

Yet, however similar these acts of quoting may be, the norms governing them today are very different. Though I've not yet found anyone who can quite express why, any qualified Hollywood lawyer would tell you there's a fundamental difference between quoting Hemingway and quoting Sam Wood's version of Hemingway. The same with music: in an opinion by perhaps one of the twentieth century's worst federal judges, Judge Kevin Thomas Duffy, the court issued "stern" sanctions against rap artists who had sampled another musical recording. Wrote the judge,

"Thou shalt not steal" has been an admonition followed since the dawn of civilization. Unfortunately, in the modern world of business this admonition is not always followed. Indeed, the defendants in this action for copyright infringement would have this court believe that stealing is rampant in the music business and, for that reason, their conduct here should be excused. The conduct of the defendants herein, however, violates not only the Seventh Commandment, but also the copyright laws of this country.[1]

Whether justified or not, the norms governing these forms of expression are far more restrictive than the norms governing text. They admit none of the freedoms that any writer takes for granted when writing a college essay, or even an essay for the *New Yorker*.

Why?

A complete answer to that question is beyond me, and therefore us, here. But we can make a start. There are obvious differences in these forms of expression. The most salient for our purposes is the democratic difference, historically, in these kinds of "writing." While writing with text is the stuff that everyone is taught to do, filmmaking and record making were, for most of the twentieth century, the stuff that professionals did. That meant it was easier to imagine a regime that required permission to quote with film and music. Such a regime was at least feasible, even if inefficient.

But what happens when writing with film (or music, or images, or every other form of "professional speech" from the twentieth century) becomes as democratic as writing with text? As Negativland's Don Joyce described to me, what happens when technology "democratiz[es] the technique and the attitude and the method [of creating] in a way that we haven't known before.... [I]n terms of collage, [what happens when] anybody can now be an artist"?[2]

What norms (and then law) will govern this kind of creativity? Should the norms we all take for granted from writing be applied to video? And music? Or should the norms from film be applied to text? Put differently: Should the "ask permission" norms be extended from film and music to text? Or should the norms of "quote freely, with attribution" spread from text to music and film?

At this point, some will resist the way I've carved up the choices. They will insist that the distinction is not between text on the one hand and film/music/images on the other. Instead, the distinction is between commercial or public presentations of text/film/music/images on the one hand, and private or noncommercial use of text/film/music/images on the other. No one expects my friend Ben to ask the Hemingway estate for permission to quote in a college essay, because no one is publishing (yet, at least) Ben's college essays. And in the same way, no one would expect Disney, for example, to have any problem with a father taking a clip from *Superman* and including it in a home movie, or with kids at a kindergarten painting Mickey Mouse on a wall.

Yet however sensible that distinction might seem, it is in fact not how the rules are being enforced just now. Again, Ben's freedom with text is the same whether it is a college essay or an article in the *New Yorker* (save perhaps if he's writing about poetry). And in fact, Disney has complained about kids at a kindergarten painting Mickey on a wall.[3] And in a setup by J. D. Lasica, every major studio except one insisted that a father has no right to include a clip of a major film in a home movie—even if that movie is never shown to anyone except the family—without paying thousands of dollars to do so.[4]

However sensible, the freedom to quote is not universal in the noncommercial sphere. Instead, those in thousand-dollar suits typically insist that "permission is vital, legally."

Nor do I believe the freedom to quote should reach universally only in the noncommercial sphere. In my view, it should reach much broader than that. But before I can hope to make that normative argument stick, we should think more carefully about why this right to quote—or as I will call it, to remix—is a critical expression of creative freedom that in a broad range of contexts, no free society should restrict.

Remix is an essential act of RW creativity. It is the expression of a freedom to take "the songs of the day or the old songs" and create with them. In Sousa's time, the creativity was performance. The selection and arrangement expressed the creative ability of the singers. In our time, the creativity reaches far beyond performance alone. But in both contexts, the critical point to recognize is that the RW creativity does not compete with or weaken the market for the creative work that gets remixed. These markets are complementary, not competitive.

That fact alone, of course, does not show that both markets shouldn't be regulated (that is, governed by rules of copyright). But as we'll see in the next part of the book, there are important reasons why we should limit the regulation of copyright in the contexts in which RW creativity is likely to flourish most. These reasons reflect more than the profit of one, albeit important, industry; instead, they reflect upon a capacity for a generation to speak.

I start with a form of RW culture that is closest to our tradition of remixing texts. From that beginning, I will build to the more significant forms of remix now emerging. In the end, my aim is to draw all these forms together to point to a kind of speech that will seem natural and familiar. And a kind of freedom that will feel inevitable.

Remixed: Text

There is a thriving RW culture for texts on the Net just now. Its scope and reach and, most important, sophistication are far beyond what anyone imagined at the Internet's birth. Through technologies not even conceived of when this system began, this RW culture for texts has built an ecology of content and an economy of reputation. There is a system now that makes an extraordinary range of initially unfiltered content understandable, and that helps the reader recognize what he should trust, and what he should question.

We can describe this system in three layers. The first is the writing itself. This has evolved through two different lives. The first of these is obscure to many; the second is the ubiquitous "blog."

The first was something called Usenet. In 1979, two computer scientists at Duke, Tom Truscott and Jim Ellis, invented a distributed messaging system that enabled messages to be passed cheaply among thousands of computers worldwide. This was Usenet. Sometimes these messages were announcementy; sometimes they were simply informational. But soon they became the location of increasingly interactive RW culture. As individuals realized they could simply hit a single button and post a comment or reply to thousands of computers worldwide, the temptation to speak could not be resisted. Usenet grew quickly, and passion around it grew quickly as well.

In 1994, a couple of lawyers changed all this. The firm Canter &

Siegel posted the first cross-group commercial message—aka spam—advertising its services. Thousands responded in anger, flaming the lawyers to get them to stop. But many others quickly copied Canter & Siegel. Other such scum quickly followed. Usenet became less and less a place where conversation could happen, and more and more a ghetto for gambling ads and other such scams (see also your e-mail in-box).[5]

Just about the time that Usenet was fading, the World Wide Web was rising. The Web's inventor, Tim Berners-Lee, was keen that the Web be a RW medium—what Benkler calls "the writable Web."[6] He pushed people developing tools to implement Web protocols to design their tools in a way that would encourage both reading and writing.[7] At first, this effort failed. The real drive for the Web, its developers thought, would be businesses and other organizations that would want to publish content to the world. RO, not RW.

But as tools to simplify HTML coding matured, Berners-Lee's idea of a RW Internet became a reality. Web-logs, or blogs, soon started to proliferate at an explosive rate. In March 2003, the best-known service for tracking blogs, Technorati, found just 100,000 blogs. Six months later, that number had grown to 1 million. A year later, more than 4 million were listed.[8] Today there are more than 100 million blogs worldwide, with more than 15 added in the time it took you to read this sentence. According to Technorati, Japanese is now the number one blogging language. And Farsi has just entered the top ten.[9]

When blogs began (and you can still see these early blogs using Brewster Kahle's "Wayback machine" at archive.org), while they expressed RW creativity (since the norm for this form of writing encouraged heavy linking and citation), their RW character was limited. Many were little more than a public diary: people (and some very weird people) posting their thoughts into an apparently empty void. Most were commentary on other public events. So the writing itself was RW, but the writing was experienced by an audience as RO.

Soon, however, in what Benkler calls the "second critical innovation of the writable Web,"[10] bloggers added a way for their audience to talk back. Comments became an integral part of blogging. Some of these comments were insightful, some were silly, some were designed simply to incite. But by adding a way to talk back, blogs changed how they were read.

This was the first layer of the Net's RW culture for text. Alone, however, this layer would be worth very little. How could you find anything of interest in this vast, undifferentiated sea of content? If you knew someone you trusted, maybe you'd read her blog. But why would you waste your time reading some random person's thoughts about anything at all?

The next two layers helped solve this problem. The first added some order to the blogosphere. It did so by adding not a taxonomy but, as Thomas Vander Wal puts it, a "folksonomy to this RW culture."[11] Tags and ranking systems, such as del.icio.us, Reddit, and Digg, enabled readers of a blog or news article to mark it for others to find or ignore. These marks added meaning to the post or story. They would help it get organized among the millions of others that were out there. Together these tools added a metalayer to the blogosphere, by providing, as *Wired* cofounder Kevin Kelly puts it, "a public annotation—like a keyword or category name that you hang on a file, Web page or picture."[12] And as readers explore the Web, users leave marks that help others understand or find the same stuff.

So, for example, if you read an article about Barack Obama, you can tag it with a short description: "Obama" or "Obama_environment." As millions of readers do the same, the system of tagging begins to impose order on the stuff tagged—even though no one has drafted a table of tags, and no one imposes any rules about the tags. You could just as well tag the Obama article "petunias," and some few petunia lovers will be disappointed as they follow the sign to this nonpetunia site. But as more and more users push the arrows in other ways, more and more follow more faithful taggers.

Tagging thus added a layer of meaning to RW content. The more tags, the more useful and significant they become. Importantly, this significance is created directly by the viewers or consumers of that culture—not by advertisers, or by any other intentional efforts at commercial promotion. This reputation and word-of-mouth technology create a competing set of meanings that get associated with any content. The tools become "powerful forces that marketers must harness," though as Don Tapscott and Anthony Williams point out, this is a force that can "just as easily spin out of control in unpredictable ways."[13]

As they add meaning to content, these tools also enable collaboration. Significance and salience are a self-conscious community activity.[14] Sites such as del.icio.us reinforce this community power by allowing users to share bookmarks, enabling "links [to] become...the basis for learning new things and making connections to new people."[15] They also change the relative power of the reader. As the reader "writes" with tags or votes, the importance of the original writing changes. A major national newspaper could have the highest-paid technology writer in the world. But what happens to that writer when it turns out that the columns read by more, and recommended by most, are written by eighteen-year-old bloggers? The *New York Times* used to have the power to say who was the most significant. A much more democratic force does that now.

The third layer of this RW culture for text is much less direct. These are tools that try to measure the significance of a conversation by counting the links that others make to the conversations. Technorati is the leader in this area so far. Its (ro)bots crawl the world of blogs, counting who links to whom or what. The company then publishes up-to-the-minute rankings and link reports, so you can post a blog entry and, minutes later, begin watching everyone who links back to that entry. Technorati says it updates its index every ten minutes.[16] With over 100 million blogs indexed, that's a very fast update.

Indices like this show the revealed preferences of the blogosphere. In almost real time, we can see who is wielding influence. And as the space matures, most interestingly, we can see that the influence of blogs is increasingly outstripping mainstream media. In the Q4 report for 2006, Technorati reported that in the 51–100 range of most popular sites on the Web, 25 percent were blogs.[17] Ten years before, 0 percent of nonprofessional content would have been among the most popular of any popular media.

These three layers, then, work together. There would be nothing without the content. But there would be too much to be useful were there only the content. So, in addition to content, content about content—tags, and recommendations—combined with tools to measure the influence of content. The whole becomes an ecosystem of reputation. Those trying to interact with culture now recognize this space as critical to delivering or understanding a message.

Many worry about this blogosphere. Some worry it is just a fad—but what fad has ever caught 100 million users before? Charlene Li reports that 33 percent of teenagers make a blog entry weekly, and 41 percent visit a social networking site daily.[18] And absolutely every major publication devotes a substantial amount of resources to making this presence as important as any.

Others worry about quality; how can bloggers match the *New York Times*? What bloggers will spend the effort necessary to get their stories right?

If the question is asked about blogs on average, then no doubt the skepticism is merited. But if the same question were asked of newspapers *on average,* then great skepticism about newspapers would be merited as well. The point with both is that we have effective tools for assessing quality. And more important, we have increasingly famous examples of blogs outdoing traditional media in delivering both quality and truth. Yochai Benkler catalogs a host of cases where bloggers did better than mainstream media in ferreting out the truth, such as uncovering the truth about Trent Lott's affection for racist statements, or the lack of veracity in Diebold's claims about its voting machines.[19] And even a cursory review of key political blogs—Instapundit or Michelle Malkin on the Right, the Daily Kos or the Huffington Post on the Left—reveals a depth and an understanding that are rare in even the best of mainstream media. The point is there's good and bad on both sides. But perhaps in the blogosphere, there are better mechanisms for determining what is good and what is bad.

The point was driven home to me in the 2004 election. That election, of course, had created public awareness about blogs, since the early front-runner, Vermont governor Howard Dean, had been created by blog culture. But as I watched the returns from that election on national television, I began to feel sorry for the "correspondents" who had to report on this pivotal election on television. In one example, one of our nation's most prominent correspondents was asked to give, as the segment was advertised, an "in-depth analysis of" voters in one particular state. When the segment began, the national desk switched to this correspondent, and he began with a question to three "average voters" from the state. He then had about thirty seconds to add his own witty insight on top of their totally inane blather about why they had voted as they did. And that was it. One minute, and zero substance, broadcast to millions across the country.

It so happened that at the same time, I was reading an "in-depth analysis" of the same state, posted on a blog. The post had been written within the previous three hours. It was chock-full of substance and insight. Timely, smart, and comprehensive—much better than the "human angle" news that is national news today, and much more reflective of the talent of a great journalist. The television reporter no doubt thought he was a journalist. But with TV tuned to the attention span of an increasingly ADD public, who can afford to be a journalist?

There's one more important dimension to the RW culture of text on the Internet: the power of advertising. In this realm, the *editorial* power of advertisers is radically smaller than in traditional media. An advertiser can choose whether and where to advertise. But advertising is still a small part of the economy of blogging, and where it is relevant, many different content sources compete, so the ability of an advertiser indirectly to control content is radically diminished.[20]

The RW Internet is an ecosystem.

Many will remain skeptical. If the quality of the average blog is so bad, what good could this RW creativity be doing? But here we need to focus upon a second aspect of RW creativity—not so much the quality of the speech it produces, but the effect it has upon the person producing the speech.

I've felt one aspect of this effect personally. I'm a law professor. For the first decade of my law professor life, I wrote with the blissful understanding that no one was reading what I wrote. That knowledge gave me great freedom. More important, whatever the three readers of my writing thought remained their own private thoughts. Law professors write for law journals. Law journals don't attach comments to the articles they publish.

Blog space is different. You can see people read your writing; if you allow, you can see their comments. The consequence of both is something you can't quite understand until you've endured it. Like eating spinach or working out, I force myself to suffer it because I know it's good for me. I've written a blog since 2002. Each entry has a link for comments. I don't screen or filter comments (save for spam). I don't require people to give their real name. The forum is open for anyone to say whatever he or she wants. And people do. Some of the comments are quite brilliant. Many add important facts I've omitted or clarify what I've misunderstood. Some commentators become regulars. One character, "Three Blind Mice," has been a regular for a long time, rarely agreeing with anything I say.

But many of the comments are as rude and abusive as language allows. There are figures—they're called "trolls"—who live for the fights they can gin up in these spaces. They behave awfully. Their arguments are (in the main) ridiculous, and they generally make comment spaces deeply unpleasant.

Other commentators find ways around these trolls. Norms like "don't feed the troll" are invoked whenever anyone takes a troll on. But there's only so much that can be done, at least so long as the forum owner (me) doesn't block certain people or force everyone to use his or her real name.

I find it insanely difficult to read these comments. Not because they're bad or mistaken, but mainly because I have very thin skin. There's a direct correlation between what I read and pain in my gut. Even unfair and mistaken criticism cuts me in ways that are just silly. If I read a bad comment before bed, I don't sleep. If I trip upon one when I'm trying to write, I can be distracted for hours. I fantasize about creating an alter ego who responds on my behalf. But I don't have the courage for even that deception. So instead, my weakness manifests itself through the practice (extraordinarily unfair to the comment writer) of sometimes not reading what others have said.

So then why do I blog all? Well, much of the time, I have no idea why I do it. But when I do, it has something to do with an ethic I believe that we all should live by. I first learned it from a judge I clerked for, Judge Richard Posner. Posner is without a doubt the most significant legal academic and federal judge of our time, and perhaps of the last hundred years. He was also the perfect judge to clerk for. Unlike the vast majority of appeals court judges, Posner writes his own opinions. The job of the clerk was simply to argue. He would give us a draft opinion, and we'd write a long memo in critique. He'd use that to redraft the opinion.

I gave Posner comments on much more than his opinions. In particular, soon after I began teaching he sent me a draft of a book, which would eventually become *Sex and Reason*. Much of the book was brilliant. But there was one part I thought ridiculous. And in a series of faxes (I was teaching in Budapest, and this was long before e-mail was generally available), I sent him increasingly outrageous comments, arguing about this section of the book.

The morning after I sent one such missive, I reread it, and was shocked by its abusive tone. I wrote a sheepish follow-up, apologizing, and saying that of course, I had endless respect for Posner, blah, blah, and blah. All that was true. So too was it true that I thought my comments were unfair. But Posner responded not by accepting my apology, but by scolding me. And not by scolding me for my abusive fax, but for my apology. "I'm surrounded by sycophants," he wrote. "The last thing in the world I need is you to filter your comments by reference to my feelings."

I was astonished by the rebuke. But from that moment on, I divided the world into those who would follow (or even recommend) Posner's practice, and those who wouldn't. And however attractive the anti-Posner pose was, I wanted to believe I could follow his ethic: Never allow, or encourage, the sycophants. Reward the critics. Not because I'd ever become a judge, or a public figure as important as Posner. But because in following his example, I would avoid the worst effects of the protected life (as a tenured professor) that I would lead.

Until the Internet, there was no good way to do this, at least if you were as insignificant as I. It's not like I could go to my local Starbucks and hold a public forum. There are people who do that in my neighborhood. Most of them have not showered for weeks. Famous people could do this, in principle. But the ethic of public appearances today, at least for Americans, militates against this sort of directness. It's rude to be critical. Indeed, if you're too critical, you're likely to be removed from the forum by men with badges.

This is not the way it is everywhere. In perhaps the most dramatic experience of democracy I've witnessed, I watched Brazil's minister of culture, Gilberto Gil, argue with a loving but critical crowd (loving his music and most of his policies, except the part that protected incumbent radio stations).[21] The forum was packed. There was no stage that separated Gil from the hundreds who huddled around to hear him. People argued directly with him. He argued back, equal to equal. The exchange was so honest that it even embarrassed John Perry Barlow, Gil's friend and fan, who stood to defend Gil against the critics.

But Gil loved the exchange. He was not embarrassed by the harshness of the criticism. His manner encouraged it. He was a democratic leader in a real (as opposed to hierarchical) democracy. He was Posner in Brazil.

For those of us who are not Posner and not Gil, the Internet is the one context that encourages the ethic of democracy that they exemplify. It is the place where all writing gets to be RW. To write in this medium is to know that anything one writes is open to debate. I used to love the conceit of a law review article—presenting its arguments as if they were proven, with little or no space provided for disagreement. I now feel guilty about participating in such a form.

All this openness is the product of a kind of democracy made real with writing. If trends continue, we're about to see this democracy made real with all writing. The publishers are going to fight the Googlezation of books. But as authors see that the most significant writing is that which is RW, they'll begin to insist that their publishers relax. In ten years, everything written that is read will be accessible on the Net—meaning not that people will be able to download copies to read on their DRM-encumbered reader but accessible in an open-access way, so that others will be able to comment on, and rate, and criticize the writing they read. This write/read is the essence of RW.

Text is just a small part of the RW culture that the Internet
is. Consider now its big, and ultimately much more significant,
sister.

Remixed: Media

For most of the Middle Ages in Europe, the elite spoke and wrote
in Latin. The masses did not. They spoke local, or vernacular, languages—what we now call French, German, and English. What
was important to the elites was thus inaccessible to the masses. The
most "important" texts were understood by only a few.

Text is today's Latin. It is through text that we elites communi-
cate (look at you, reading this book). For the masses, however, most
information is gathered through other forms of media: TV, film,
music, and music video. These forms of "writing" are the vernacular of today. They are the kinds of "writing" that matters most to
most. Nielsen Media Research, for example, reports that the average TV is left on for 8.25 hours a day, "more than an hour longer
than a decade ago."[22] The average American watches that average
TV about 4.5 hours a day.[23] If you count other forms of media—
including radio, the Web, and cell phones—the number doubles.[24]
In 2006, the U.S. Bureau of the Census estimated that "American
adults and teens will spend nearly five months" in 2007 consuming media.[25] These statistics compare with falling numbers for text.
Everything is captured in this snapshot of generations:

Individuals age 75 and over averaged 1.4 hours of reading per
weekend day and 0.2 hour (12 minutes) playing games or using a

computer for leisure. Conversely, individuals ages 15 to 19 read for
an average of 0.1 hour (7 minutes) per weekend day and spent 1.0
hour playing games or using a computer for leisure.[26]

It is no surprise, then, that these other forms of "creating" are
becoming an increasingly dominant form of "writing." The Internet didn't make these other forms of "writing" (what I will call
simply "media") significant. But the Internet and digital technologies opened these media to the masses. Using the tools of digital
technology—even the simplest tools, bundled into the most innovative modern operating systems—anyone can begin to "write" using
images, or music, or video. And using the facilities of a free digital
network, anyone can share that writing with anyone else. As with
RW text, an ecology of RW media is developing. It is younger than
the ecology of RW texts. But it is growing more quickly, and its
appeal is much broader.[27]

These RW media look very much like Ben's writing with text.
They remix, or quote, a wide range of "texts" to produce something
new. These quotes, however, happen at different layers. Unlike text,
where the quotes follow in a single line—such as here, where the
sentence explains, "and then a quote gets added"—remixed media
may quote sounds over images, or video over text, or text over
sounds. The quotes thus get mixed together. The mix produces the
new creative work—the "remix."

These remixes can be simple or they can be insanely complex.
At one end, think about a home movie, splicing a scene from *Superman* into the middle. At the other end, there are new forms of art
being generated by virtuosic remixing of images and video with
found and remade audio. Think again about Girl Talk, remixing

between 200 and 250 samples from 167 artists in a single CD. This is not simply copying. Sounds are being used like paint on a palette. But all the paint has been scratched off of other paintings.

So how should we think about it? What does it mean, exactly?

However complex, in its essence remix is, as Negativland's Don Joyce described to me, "just collage." Collage, as he explained,

> [e]merged with the invention of photography. Very shortly after it was invented... you started seeing these sort of joking postcards that were photo composites. There would be a horse-drawn wagon with a cucumber in the back the size of a house. Things like that. Just little joking composite photograph things. That impressed painters at the time right away.

But collage with physical objects is difficult to do well and expensive to spread broadly. Those barriers either kept many away from this form of expression, or channeled collage into media that could be remixed cheaply. As Mark Hosler of Negativland described to me, explaining his choice to work with audio,

> I realized that you could get a hold of some four-track reel-to-reel for not that much money and have it at home and actually play around with it and experiment and try out stuff. But with film, you couldn't do that. It was too expensive.... So that... drove me... to pick a medium where we could actually control what we · were doing with a small number of people, to pull something off and make some finished thing to get it out there.[28]

With digital objects, however, the opportunity for wide-scale collage is very different. "Now," as filmmaker Johan Söderberg explained, "you can do [video remix] almost for free on your own computer."[29] This means more people can create in this way, which means that many more do. The images or sounds are taken from the tokens of culture, whether digital or analog. The tokens are "blaring at us all the time," as Don Joyce put it to me: "We are barraged" by expression intended originally as simply RO. Negativland's Mark Hosler:

> When you turn around 360 degrees, how many different ads or logos will you see somewhere in your space? [O]n your car, on your wristwatch, on a billboard. If you walk into any grocery store or restaurant or anywhere to shop, there's always a soundtrack playing. There's always...media. There's ads. There's magazines everywhere....[I]t's the world we live in. It's the landscape around us.

This "barrage" thus becomes a source.[30] As Johan Söderberg says, "To me, it is just like cooking. In your cupboard in your kitchen you have lots of different things and you try to connect different tastes together to create something interesting."

The remix artist does the same thing with bits of culture found in his digital cupboard.

My favorites among the remixes I've seen are all cases in which the mix delivers a message more powerfully than any original alone could, and certainly more than words alone could.

For example, a remix by Jonathan McIntosh begins with a scene from *The Matrix*, in which Agent Smith asks, "Do you ever get the feeling you're living in a virtual reality dream world? Fabricated to enslave your mind?" The scene then fades to a series of unbelievable war images from the Fox News Channel—a news organization that arguably makes people less aware of the facts than they were before watching it.[31] Toward the end, the standard announcer voice says, "But there is another sound: the sound of good will." On the screen is an image of Geraldo Rivera, somewhere in Afghanistan. For about four seconds, he stands there silently, with the wind rushing in the background. (I can always measure the quickness of my audience by how long it takes for people to get the joke: "the sound of good will" = silence). The clip closes with a fast series of cuts to more Fox images, and then a final clip from an ad for the film that opened McIntosh's remix: "The Matrix Has You."

Or consider the work of Sim Sadler, video artist and filmmaker. My favorite of his is called "Hard Working George." It builds exclusively from a video of George Bush in one of his 2004 debates with John Kerry. Again and again, Sadler clips places where Bush says, essentially, "it's hard work." Here's the transcript:

Sir, in answer to your question I just know how this world works. I see on TV screens how hard it is. We're making progress; it is hard work. You know, it's hard work. It's hard work. A lot of really great people working hard, they can do the hard work. That's what distinguishes us from the enemy. And it's hard work, but it's necessary work and that's essential, but again I want to tell the American people it's hard work. It is hard work. It's hard work. There is no doubt in my mind that it is necessary work. I understand how hard it is, that's my job. No doubt about it, it's tough. It's hard work which I really want to do, but I would hope I never have to—nothing wrong with that. But again I repeat to my fellow citizens, we're making progress. We're making progress there. I reject this notion. It's ludicrous. It is hard work. It's hard

work. That's the plan for victory and that is the best way. What I
said was it's hard work and I made that very clear.

Usually, the audience breaks into uncontrolled laughter at "I
would hope I never have to—nothing wrong with that," so people
don't hear the rest of the clip. But by the end, the filter Sadler has
imposed lets us understand Bush's message better.

Some look at this clip and say, "See, this shows anything can
be remixed to make a false impression of the target." But in fact,
the "not working hard" works as well as it does precisely because
it is well known that at least before 9/11, Bush was an extremely
remote president, on vacation 42 percent of his first eight months
in office.[32] The success of the clip thus comes from building upon
what we already know. It is powerful because it makes Bush himself say what we know is true about him. The same line wouldn't
have worked with Clinton, or Bill Gates. Whatever you want to say
about them, no one thinks they don't work hard.

My favorite of all these favorites, however, is still a clip in a series
called "Read My Lips," created by Söderberg. Söderberg is an artist,
director, and professional video editor. He has edited music videos for
Robbie Williams and Madonna and, as he put it, "all kinds of pop
stars." He also has an Internet TV site—soderberg.tv—that carries
all his own work. That work stretches back almost twenty years.

"Read My Lips" is a series Söderberg made for a Swedish company called Atmo, in which famous people are lip-synched with
music or other people's words. They all are extraordinarily funny
(though you can't see all of them anymore because one, which mixed
Hitler with the song "Born to Be Alive," resulted in a lawsuit).

The best of these (in my view at least) is a love song with Tony

Blair and George Bush. The sound track for the video is Lionel Richie's "Endless Love." Remember the words "My love, there's only you in my life." The visuals are images of Bush and Blair. Through careful editing, Söderberg lip-synchs Bush singing the male part and Blair singing the female part. The execution is almost perfect. The message couldn't be more powerful: an emasculated Britain, as captured in the puppy love of its leader for Bush.

The obvious point is that a remix like this can't help but make its argument, at least in our culture, far more effectively than could words. (By "effectively," I mean that it delivers its message success-fully to a wide range of viewers.) For anyone who has lived in our era, a mix of images and sounds makes its point far more power-fully than any eight-hundred-word essay in the *New York Times* could. No one can deny the power of this clip, even Bush and Blair supporters, again in part because it trades upon a truth we all— including Bush and Blair supporters—recognize as true. It doesn't assert the truth. It shows it. And once it is shown, no one can escape its mimetic effect. This video is a virus; once it enters your brain, you can't think about Bush and Blair in the same way again.

But why, as I'm asked over and over again, can't the remixer simply make his own content? Why is it important to select a drumbeat from a certain Beatles recording? Or a Warhol image? Why not simply record your own drumbeat? Or paint your own painting?

The answer to these questions is not hard if we focus again upon why these tokens have meaning. Their meaning comes not from the content of what they say; it comes from the reference, which is expressible only if it is the original that gets used. Images or sounds collected from real-world examples become "paint on a palette."

And it is this "cultural reference," as coder and remix artist Victor Stone explained, that "has emotional meaning to people.... When you hear four notes of the Beatles' 'Revolution,' it means something."[33] When you "mix these symbolic things together" with something new, you create, as Söderberg put it, "something new that didn't exist before."

The band Negativland has been making remixes using "found culture"—collected recordings of RO culture—for more than twenty-five years. As I described at the start, they first became (in)famous when they were the target of legal action brought by Casey Kasem and the band U2 after Negativland released a mash-up of Casey Kasem's introduction of U2 on his Top 40 show. So why couldn't Negativland simply have used something original? Why couldn't they rerecord the clip with an actor? Hosler explained:

> We could have taken these tapes we got of Casey Kasem and hired someone who imitated Casey Kasem, you know, and had him do a dramatic re-creation. Why did we have to use the actual original...the actual thing? Well, it's because the actual thing has a power about it. It has an aura. It has a magic to it. And that's what inspires the work.

Likewise with their remarkable, if remarkably irreverent, film, *The Mashin' of the Christ*. This five-minute movie is made from remixing the scores of movies made throughout history about Jesus' crucifixion. The audio behind these images is a revivalist preacher who repeatedly says (during the first minute), "Christianity is stupid." The film then transitions at about a minute and a half when the preacher says, "Communism is good." The first quote aligns Christians, at least, against the film. But the second then reverses that feeling, as the film might also be seen as a criticism of Communism. As Hosler explained the work:

> *The Mashin' of the Christ* just came out of an idle thought that crossed my mind one day when I was flipping around on Amazon .com. I thought, "How many movies have been made about the life of Jesus, anyway?" I came up with thirty or forty of them and I started thinking about [how] every one of those films has similar sequences of Jesus being beaten, flogged, whipped, abused. There's always a shot where he's carrying the cross and he stumbles and he falls. And it just occurred to me...I thought that would make an interesting montage of stuff.

This montage's point could not have been made by simply shooting crucifixion film number forty-one.

The Significance of Remix

I've described what I mean by remix by describing a bit of its practice. Whether text or beyond text, remix is collage; it comes from combining elements of RO culture; it succeeds by leveraging the meaning created by the reference to build something new.

But why should anyone care about whether remix flourishes, or even exists? What does anyone gain, beyond a cheap laugh? What does a society gain, beyond angry famous people?

There are two goods that remix creates, at least for us, or for our kids, at least now. One is the good of community. The other is education.

COMMUNITY

Remixes happen within a community of remixers. In the digital age, that community can be spread around the world. Members of that community create in part for one another. They are showing one another how they can create, as kids on a skateboard are showing their friends how they can create. That showing is valuable, even when the stuff produced is not.

Consider, for example, the community creating anime music videos (AMV). Anime are the Japanese cartoons that swept America a few years ago. AMVs are (typically) created by remixing images from these cartoons with a music track or the track from a movie trailer. Each video can take between fifty and four hundred hours to create. There are literally thousands that are shared non-commercially at the leading site, animemusicvideos.org.

The aim of these creators is in part to learn. It is in part to show off. It is in part to create works that are strikingly beautiful. The work is extremely difficult to do well. Anyone who does it well also has the talent to do well in the creative industries. This fact has not been lost on industry, or universities training kids for industry. After I described AMVs at one talk, a father approached me with tears in his eyes. "You don't know how important this stuff is," he told me. "My kid couldn't get into any university. He then showed them his AMVs, and now he's at one of the best design schools in America."

AMVs are peculiarly American—or, though they build upon

Japanese anime, they are not particularly Japanese. This is not because Japanese kids are not remixers. To the contrary, Japanese culture encourages this remixing from a much younger age, and much more broadly. According to cultural anthropologist Mimi Ito,

> Japanese media have really been at the forefront of pushing recombinant and user-driven content starting with very young children. If you consider things like *Pokémon* and *Yu-Gi-Oh!* as examples of these kinds of more fannish forms of media engagement, the base of it is very broad in Japan, probably much broader than in the U.S. Something like *Pokémon* or *Yu-Gi-Oh!* reached a saturation point of nearly 100 percent within kids' cultures in Japan.[34]

But the difference between cultures is not just about saturation. Henry Jenkins quotes education professors David Buckingham and Julia Sefton-Green, "*Pokémon* is something you do, not just something you read or watch or consume," and continues:

> There are several hundred different *Pokémon*, each with multiple evolutionary forms and a complex set of rivalries and attachments. There is no one text where one can go to get the information about these various species; rather, the child assembles what they know about the *Pokémon* from various media with the result that each child knows something his or her friends do not and thus has a chance to share this expertise with others.[35]

"Every person," Ito explains, thus "has a personalized set of *Pokémon*. That is very different from [American media, which are] asking kids to identify with a single character."

Pokémon is just a single example of a common practice in Japan.

This more common practice pushes "kids to develop more persona lives, and remix-oriented pathways to the content." Kids in the second and third grades, for example, will all

> carry around just a little sketchbook . . . with drawings of manga [cartoon] characters in them. That's what [Japanese] kids do. Then by fourth or fifth grade there are certain kids that get known to be good at drawing and then they actually start making their original stories. Then at some point there needs to be an induction into the whole *doujinshi* scene, which is its own subculture. That usually happens through knowing an older kid who's involved in that.

American kids have it different. The focus is not: "Here's something, do something with it." The focus is instead: "Here's something, buy it." "The U.S. has a stronger cultural investment in the idea of childhood innocence," Ito explains, "and it also has a more protectionist view with respect to media content." And this "protectionism" extends into schooling as well. "Entertainment" is separate from "education." So any skill learned in this "remix culture" is "constructed oppositionally to academic achievement." Thus, while "remix culture" flourishes with adult-oriented media in the United States, "there's still a lot of resistance to media that are coded as children's media being really fully [integrated] into that space."

Yet the passion for remix is growing in American kids, and AMVs are one important example. Ito has been studying these AMV creators, getting a "sense of their trajectories" as creators. At what moment, she is trying to understand, does "a fan see [himself] as a media producer and not just a consumer"? And what was the experience (given it was certainly not formal education) that led them to this form of expression?

Ito's results are not complete, but certain patterns are clear. "A very high proportion of kids who engage in remix culture," for example, "have had experience with interactive gaming formats." "The AMV scene is dominated by middle-class white men"—in contrast to the most famous remixers in recent Japanese history, the "working-class girls" who produced *doujinshi*. Most "have a day job or are full-time students but...have an incredibly active amateur life.... [They] see themselves as producers and participants in a culture and not just recipients of it." That participation happens with others. They form the community. That community supports itself.

EDUCATION

A second value in remix extends beyond the value of a community. Remix is also and often, as Mimi Ito describes, a strategy to excite "interest-based learning." As the name suggests, interest-based learning is the learning driven by found interests. When kids get to do work that they feel passionate about, kids (and, for that matter, adults) learn more and learn more effectively.

I wrote about this in an earlier book, *Free Culture*. There I described the work of Elizabeth Daley and Stephanie Barish, both of whom were working with kids in inner-city schools. By giving these kids basic media literacy, they saw classes of students who before could not retain their focus for a single period now spending every free moment of every hour the school was open editing and perfecting video about their lives, or about stories they wanted to tell.

Others have seen the same success grow from using remix media to teach. At the University of Houston—a school where a high percentage of the students don't speak English as their first language—the Digital Storytelling project has produced an extraordinary range of historical videos, created by students who research the story carefully, and select from archives of images and sounds the mix that best conveys the argument they want their video to make.

As Henry Jenkins notes, "[M]any adults worry that these kids are 'copying' preexisting media content rather than creating their own original works."[36] But as Jenkins rightly responds, "More and more literacy experts are recognizing that enacting, reciting, and appropriating elements from preexisting stories is a valuable and organic part of the process by which children develop cultural literacy."[37] Parents should instead, Jenkins argues, "think about their [kids'] appropriations as a kind of apprenticeship."[38] They learn by remixing. Indeed, they learn more about the form of expression they remix than if they simply made that expression directly.

This is not to say, of course, that however they do this remix, they're doing something good. There's good and bad remix, as there's good and bad writing. But just as bad writing is not an argument against writing, bad remix is not an argument against remix. Instead, in both cases, poor work is an argument for better education. As Hosler put it to me:

Every high school in America needs to have a course in media literacy. We're buried in this stuff. We're breathing it. We're drinking it constantly. It's 24/7 news and information and pop culture. . . . If you're trying to educate kids to think critically about history and society and culture, you've got to be encouraging them to be thoughtful and critical about media and information and advertising.

Doing something with the culture, remixing it, is one way to learn.

The Old in the New

To many, my description of remix will sound like something very new. In one sense it is. But in a different, perhaps more fundamental sense, we also need to see that there's nothing essentially new in remix. Or put differently, the interesting part of remix isn't something new. All that's new is the technique and the ease with which the product of that technique can be shared. That ease invites a wider community to participate; it makes participation more compelling. But the creative act that is being engaged in is not significantly different from the act Sousa described when he recalled the "young people together singing the songs of the day or the old songs."

For as I've argued, remix with "media" is just the same sort of stuff that we've always done with words. It is how Ben wrote. It is how lawyers argue. It is how we all talk all the time. We don't notice it as such, because this text-based remix, whether in writing or conversation, is as common as dust. We take its freedoms for granted. We all expect that we can quote, or incorporate, other people's words into what we write or say. And so we do quote, or incorporate, or remix what others have said.

The same with "media." Remixed media succeed when they show others something new; they fail when they are trite or derivative. Like a great essay or a funny joke, a remix draws upon the work of others in order to do new work. It is great writing without words. It is creativity supported by a new technology.

Yet though this remix is not new, for most of our history it was silenced. Not by a censor, or by evil capitalists, or even by good capitalists. It was silenced because the economics of speaking in this different way made this speaking impossible, at least for most. If in 1968 you wanted to capture the latest Walter Cronkite news program and remix it with the Beatles, and then share it with your ten thousand best friends, what blocked you was not the law. What blocked you was that the production costs alone would have been in the tens of thousands of dollars.

Digital technologies have now removed that economic censor. The ways and reach of speech are now greater. More people can use a wider set of tools to express ideas and emotions differently. More can, and so more will, at least until the law effectively blocks it.

FIVE

CULTURES COMPARED

've described two cultures and two kinds of creativity. One (RO) is fueled by professionals. The other (RW) is fueled by both professionals and amateurs. Both have been critical to the development of culture. Both will be spread by the maturing of digital technologies. But though I believe both will grow in the digital age, there are still important differences between them. In this brief interlude, consider a few of these differences. Then, before we turn to perhaps the most interesting development, consider some lessons that understanding these two cultures can teach.

Differences in Value—and "Values"

These two cultures embody different values.

RO culture speaks of professionalism. Its tokens of culture demand a certain respect. They offer themselves as authority. They teach, but not by inviting questions. Or if they invite questions, they direct the questions to someone other than the speaker. Or performer. Or creator.

This form of culture is critically important, both to the spread of culture and to the spread of knowledge. There are places where authority is required: No one should want Congress's laws on a wiki. Or instructions for administering medication. Or the flight plan of a commercial airliner.

So too is RO culture central to the growth of the arts. The ability to channel the commercial return from music or film has allowed many people to create who otherwise could not. This is the proper function of copyright law, and its only good justification. Where we can see that creativity would be hindered by the absence of this special privilege, the privilege makes sense.

And finally, RO culture makes possible an integrity to expression that, for some at least, is crucial. Artists want their expression framed just as they intend it. RO culture gives them that freedom. Doctors or pharmaceutical companies want to assure that instructions or medical explanations are not translated by just anyone. Control here is important, and not at all evil. Again, where it gives us something we otherwise wouldn't have—artistic expression or quality assurance—control can be good.[1]

RW culture extends itself differently. It touches social life differently. It gives the audience something more. Or better, it asks something more of the audience. It is offered as a draft. It invites a response. In a culture in which it is common, its citizens develop a kind of knowledge that empowers as much as it informs or entertains.

I see this difference directly in my life as a teacher. When students come to law school, most come from an essentially RO education. For four years (or more), they've sat in large lecture halls, with a professor at the front essentially reading the same lectures she's given year after year after year. "Any questions?" usually elicits points of order, not substance. "Do we have to read chapter 5?" "Will the subjunctive be on the exam?"

Maybe that's the appropriate way to teach most undergraduate courses. But the best legal education is radically different. The law school classroom is an argument. The professor provides the source for that argument. The class is a forum within which that argument happens. Students don't listen to lectures. They help make the lecture. They are asked questions; those questions frame a discussion. The structure demands that they create as they participate in the discussion.

People who know little about how the law works are puzzled—sometimes terrified—when they see that this is how we train professionals. The model of biochemistry is more attractive to them: "Here's a list of things to memorize. Do it." But the law is not a list of statutes. The law is a way of speaking and thinking and, most important, an ethic. Every lawyer must feel responsible for the law he or she helps make. For within the American system, at least, the law is made as it is practiced. How it is made depends upon the values its practitioners share.

This form of education teaches responsibility as well as the subject. It develops an ethic as well as knowledge about a particular field. And it expresses a sometimes profound respect for its students: from their very first week in law school, they are part of the conversation that law is. Their views are respected—at least so long as they place them within the frame of the law's conversation.

All of us believe this at some level. We all believe that writing has its own ethic, and that it imposes that ethic on the writer and the thing written. Those of us who object to judges who delegate opinion writing to their law clerks do so not so much because we want judges who are bettered by becoming better writers, but

instead because we want opinions that bear the mark of the con-
straint that comes from the responsibility of writing. Creating is a
responsibility. Only by practicing it can you learn it.

Those who study juries say they have much the same effect on
the citizens who serve on them. Jurors are given evidence. They
act on it. As they deliberate, they recognize they've been credited
with (sometimes) extraordinary power. That (sometimes) wakes
them up. They understand they have a responsibility that reaches
far beyond their ordinary lives. That makes them think and act
differently—even after they have rendered their verdict.

These examples bias me. Of course I think reading is important. Of course it is "fundamental." But humans reach far beyond
the fundamental. And as I watch my kids grow, the part I cher-
ish the most is not their reading. It is their writing. Since my oldest (now five) was two, we have told him "monster man" stories.
Watching his rapt attention at every twist in these totally on-the-fly
made-up stories was a kick. But the moment he first objected to
a particular shift in the plot, and offered his own, was one of the
coolest moments of my life. What we want to see in our kids is their
will. What we want to inspire is a will that constructs well.

I want to see this capacity expressed not just in words. I want to
see it expressed in every form of cultural meaning. I want to watch
as he changes the ending to a song he almost loves. Or adds a character to a movie that he deeply identifies with. Or paints a picture to
express an idea that before was only latent. I want this RW capacity
in him, generalized. I want him to be the sort of person who can
create by remaking.

This then is the first difference between RO and RW cultures.
One emphasizes learning. The other emphasizes learning by speaking. One preserves its integrity. The other teaches integrity. One

emphasizes a hierarchy. The other hides the hierarchy. No one would argue we need less of the first, RO culture. But anyone who has really seen it believes we need much more of the second.

Differences in Value (As in $)

The story so far emphasizes values with a capital V. Lefties who promote social, educational, and democratic ideals would very much like these values. They speak to the sort of stuff Lefties are supposed to like.

But there are more reasons to support RW culture than the fact that a bunch of us tree huggers would like it.

For as well as promoting certain values that at least some of us find important, the RW culture also promotes economic value.

To see just why, think for second about the devices necessary to make RO culture work. You see them everywhere. They are smaller and smaller, and cheaper and cheaper. As bandwidth grows, they more efficiently grab content, which they then enable you to consume. Their brilliance has grabbed me. I don't watch "television" anymore. But I do watch my iPod connected to a screen more and more. I spend way too much money on this. Apple hopes others will spend way too much, too.

But the economic value in this consumption is tiny compared with the economic potential of consumer-generated content. Think of all the devices you need to make that home movie of your kid as Superman—the camera, the microphone, the hard disk to store 500 gigabytes of takes, the fast computer to make the rendering bearable. And then think about the bandwidth you'll need to share this creativity with your family and friends.

This is a point that has been made for some time, perhaps never better than in Andrew Odlyzko's essay "Content Is Not King":[2] despite the rhetoric of the content industry, the most valuable contribution to our economy comes from connectivity, not content. Content is the ginger in gingerbread—important, no doubt, but nothing like the most valuable component in the mix.

People on the Right need to recognize this. As I've watched the debate about copyright develop, I've been astonished by how quickly those on the Right have been captured by the content industry. Maybe I'm astonished because, as Stewart Baker chided me in a review of *Free Culture,* liberals like me spend too much time talking to liberals like me rather than to conservatives. As Baker wrote in his review,

> Viewed up close, copyright bears little resemblance to the kinds of property that conservatives value. Instead, it looks like a constantly expanding government program run for the benefit of a noisy, well-organized interest group—like Superfund, say, or dairy subsidies, except that the benefits go not to endangered homeowners or hard-working farmers but to the likes of Barbra Streisand and Eminem....Copyright is a trial lawyer's dream— a regulatory program enforced by private lawsuits where the plaintiffs have all the advantages, from injury-free damages awards to liability doctrines that extract damages from anyone who was in the neighborhood when an infringement occurred.[3]

Baker's is a great point. Let me add to it: As conservative economists have taught us again and again, value in an economy is least likely to come from state-protected monopolies. It is most likely to be generated by competition. There is a huge and vibrant economy

of competition that drives technology in our economy. The monopoly rights we call copyright are constraints on that competition. I
believe those constraints are necessary. But as with every necessary
evil, they should be as limited as possible. We should provide protection from competition only where there is a very good reason to
protect.

My point of course is not that we can or should simply sacrifice
RO culture to enable RW. Instead, the opposite: in protecting RO
culture, we shouldn't kill off the potential for RW.

Differences in Value
(As in "Is It Any Good?")

In June 2007, the backlash against RW culture was born. In a short
and cleverly written book titled *The Cult of the Amateur,* Andrew
Keen, a writer and failed Internet entrepreneur, launched a full-
scale attack on precisely the culture that I am praising. The core of
his attack was that "amateur culture" is killing "our culture." The
growth of this kind of creativity will eventually destroy much that
we think of as "good" in society. "Not a day goes by without some
new revelation that calls into question the reliability, accuracy, and
truth of the information we get from the Internet," Keen writes.[4]
And in response to all the free stuff the Internet offers, Keen is
quite worried: "What is free," he warns, "is actually costing us a
fortune."[5] Wikipedia, for example, "is almost single-handedly killing the traditional information business."[6] And the "democratization" that I praise "is," he argues, "undermining truth, souring civic
discourse, and belittling expertise, experience and talent."[7]

There is more than a bit of self-parody in Keen's book. For though the book attacks the Internet for its sloppiness and error, it itself is riddled with sloppy errors.[8] (Here's a favorite: "Every defunct record label, or laid-off newspaper reporter, or bankrupt independent bookstore is a consequence of 'free' user-generated Internet content—from craigslist's free advertising, to YouTube's free music videos, to Wikipedia's free information."[9] *"Every"*? Wow!)

Yet even if we ignore Keen, his point can't be ignored. There are many who have expressed a similar fear about the dangers that they perceive in this latest form of creativity.

My first exposure to this skepticism was at a conference at New York University about "fair use." The Comedies of Fair U$e conference was filled with artists and creators demonstrating precisely the creativity I've been praising. But in the middle of this conference, Charles Sims, a lawyer with the firm of Proskauer Rose, pleaded with the young creators to turn away from their "derivative" form of creativity. They should focus, Sims argued, on something really challenging—"original creativity." Sims said:

> I can't say strongly enough that I think what Larry is really fundamentally focused on...[—] this parasitic reuse [—] is such a terrible diversion of young people's talent....I think that if you have young film people you should be encouraging them to make their films and not to simply spend all of their time diddling around with footage that other people have made at great expense, to create stuff that's not very interesting. There's a fundamental failure of imagination....
>
> I'm saying, that as members of the academy, to encourage young people to think that instead of creating out of their own souls and

their own talents to simply reuse what's available off the streets to them, is underselling the talents that young people have.[10]

There are a number of layers to this form of criticism. We should be careful to tease them apart.

Most obvious is the criticism that the work on average is more than "creative" work. There's no comparing ten minutes produced by J. J. Abrams and ten minutes from any of the stuff that passes for video production on YouTube. Remix is just "crap." This criticism is certainly true. The vast majority of remix, like the vast majority of home movies, or consumer photographs, or singing in the shower, or blogs, is just crap. Most of these products are silly or derivative, a waste of even the creator's time, let alone the consumer's.

But I never quite get what those who raise this criticism think follows from the point. I was once a student at the Goethe Institute in Berlin. After a week of our monthlong intensive German course, I asked the teacher why we weren't encouraged to speak more. "Your German is really quite awful just now," she told me. "You would all make terrible mistakes if you spoke, so I think it best if you just listen." No doubt her assessment was right. But, amazingly for a language teacher at the Goethe Institute, she was simply missing the point.

So too do critics who argue that the vast majority of remix is bad. Think again about blogs. The value of blogs is not that I'm likely to find a comment that surpasses the very best of the *New York Times*. I'm not. But that's not the point. Blogs are valuable because they give millions the opportunity to express their ideas in writing. And with a practice of writing comes a certain important integrity. A culture filled with bloggers thinks differently about politics or public affairs, if only because more have been forced through the discipline of showing in writing why A leads to B.

If this point weren't true, why would we teach our kids how to write? Given that the vast majority will never write anything more than an e-mail or a shopping list, why is it important to torture them with creative writing essays? For again, like the Internet, the vast majority of what students write is just crap (trust me on this one). What reason is there for wasting their time (and worse, mine) to produce such garbage?

(That's a rhetorical question. I suspect you see the point.)

A second layer to this criticism is more relevant but even less true. Some criticize what I'm calling "remix" by arguing there's no there there. This is Sims's real complaint. Even granted that most is crap, even the best, in his view, is a waste, "a fundamental failure of imagination."

But anyone who thinks remixes or mash-ups are neither original nor creative has very little idea about how they are made or what makes them great. It takes extraordinary knowledge about a culture to remix it well. The artist or student training to do it well learns far more about his past than one committed to this (in my view, hopelessly naive) view about "original creativity." And perhaps more important, the audience is constantly looking for more as the audience reads what the remixer has written. Knowing that the song is a mix that could draw upon all that went before, each second is an invitation to understand the links that were drawn— their meaning, the reason they were included. The form makes demands on the audience; they return the demands in kind.

This point links directly with an argument advanced by Steven Johnson in his fantastic book, *Everything Bad Is Good for You*.[11]

Aiming to rebut the view that television has become "brain dead," Johnson argues that TV has in fact become more rich and complex over time, not less. The reason relates in part to technology. As people collect not only television sets but DVD players, producers of television programming have a strong incentive to give their audience an interest in after-broadcast sales. A show maximizes its revenue when there's a postbroadcast demand for DVDs or for reruns.

So how do you create that demand? One way is through complexity. As Johnson demonstrates, the most successful television shows have multiplied the number of plot lines running through them. And though the shows are always understandable at one viewing, few viewers would understand everything going on in every show. The fan thus has a reason to watch it again—which means, buy the DVD or tune in to reruns. Complexity thus drives follow-on consumption. Henry Jenkins makes a related point about movies:

> The old Hollywood system depended on redundancy to ensure that viewers could follow the plot at all times, even if they were distracted or went out to the lobby for a popcorn refill during a crucial scene. The new Hollywood demands that we keep our eyes on the road at all times, and that we do research before we arrive at the theater.[12]

Obviously, this idea can be taken too far. The story can't be totally inaccessible. There must be some payoff for watching the first time. But the key is to make the first time, and the next ten times, worth it. To make once necessary but not enough.

This strategy is not new with television. Think about the great nineteenth-century novels. When an author such as Dickens wrote his novels in serial form (where each chapter originally appears as an installment in a magazine, published before the full novel was completed),[13] his goal was to attract people to each installment, and then to draw them back to the finished book. Dickens (and many others) did this by writing very intricate stories that repaid many rereadings.

Remix is doing the same thing with other forms of culture. Like the work of a great classical composer (Mahler and Beethoven fit here), the best remix is compelling both the first time and the hundredth time. Indeed, it is only by the hundredth time that one begins to understand it enough for it to make sense. There needs to be a strong enough raw pull to get the listener to that hundredth hearing (something Arnold Schoenberg never quite got). But once you're hooked, you don't fight to get free. You listen again and again, each time hoping to understand more.

Even with nonclassical music, this isn't completely new. Why does one listen to Bob Dylan a thousand times, or to the ballads of Jeff Tweedy again and again? It's not just the music that compels the repeated listening. In this form, the melody is best if it's simple and compelling. Instead, it is the poetry that the melody illustrates. You listen again and again because each time you understand the poetry differently—more, and more fit to the context.

Thus the argument in favor of remix—the essential art of the RW culture—is not simply the negative: what harm are they doing? The argument for me is strongly positive: I want my kids to listen to SilviaO's remix of fourstones' latest work—a thousand times I want them to listen. Because that listening is active, and engaged, far more than the brain-dead melodies or lyrics of a Britney Spears. Her work draws on nothing, save the forbidden and erotic. In this, it may be, in Sims's view, perfectly original. Yet it is also totally derivative, and deeply disrespectful of the tradition from which she comes. You pay respect to tradition by incorporating it. But you make the tradition compelling by doing so in a way that makes everyone want to understand more. As the novelist Jonathan Lethem puts it, "What we want from every artist is that they surprise us and show us something unprecedented. But . . . this act is itself innately supported by response, appropriate, imitation."[14]

And then there's one final layer of this criticism that in the end annoys me the most. Maybe some of this work is okay, this criticism says, and some of it is even quite good. But none of it is as good as the greats of [you pick the time]. Our culture, the story goes, is collapsing. There are no standards anymore. There is no quality. Taste and art are wasting away.

Every generation has had the experience of an older generation insisting that the new is degraded, that only the old is great. Didn't that experience teach us something? As Ithiel de Sola Pool said almost two decades ago, "Each generation sees as its culture those values and practices with which it grew up. Its hallowed traditions are those it learned in childhood."[15] Of course the new is inferior to the past. How else could it be progress?

But even accepting this critique, what should be done about it? If our culture is collapsing because millions of people are choosing to watch or create stuff that the critic doesn't like, should the government intervene on behalf of the critic? I'd be the first to admit that the state has a role in regulating society, and I'm even willing to admit that sometimes the state has a role in regulating speech. Copyright law, after all, is a regulation of speech, and justified if it produces incentives to create speech that otherwise wouldn't be created.[16]

But none of these justifications for state regulation could ever support the idea that we intervene to suppress a form of "culture" that some elite believes is not good enough. Subsidies are one thing. Prohibition is something radically different.

Maybe, however, the most effective response to this criticism is one that Victor Stone, architect of ccMixter, offered me: "You know... this discussion will be over in ten or twenty years. As the boomers die out, and they get over themselves by dying, the generation that follows... just doesn't care about this discussion. They just assume that remixing is part of music, and it's part of the process, and that's it."

And they'll defend it, at least until a new form of creativity comes along that they try to stop. We all become our parents.

Differences in Law
(As in "Is It Allowed?")

American copyright law regulates (at least potentially)[17] any creative work produced after 1923, for a maximum term of life of the author plus seventy years, or ninety-five years for corporate work or work created before 1978. This regulation relates differently to RO and RW cultures. Put simply, current copyright law supports the practices of the RO culture and opposes the practices of the RW culture. Or again, as the law is architected just now, it clearly favors one kind of culture over another.

Consider first copyright's relationship to RO culture. As I've described, the essence of RO is that the user, or consumer, is given the permission to consume the culture he purchases. He has no legal permission beyond that permission to consume. During the analog history of RO culture, he had no easy technical capacity beyond that capacity to consume.

Digital technologies changed the technical capacity. In its first version, digital technologies gave users almost unlimited technical capacity to mix and remix RO culture. But "can" does not imply "may." While the tools enabled users to do with RO culture as they wished, the law did not grant users of RO culture the permission to do as they wished. Instead, as applied to a world where each use of culture is a copy, RO culture required the permission of copyright owners before RO culture could be remade.

This gap between what the law permitted and what the technology allowed could have been closed either by changing the law or by changing the technology. The past five years have seen changes in both—but both aiming to strengthen the control over content, not weaken it. As RO culture has evolved in the digital world, technologies have given the copyright owner an ever-increasing opportunity to control precisely how copyrighted content is consumed. At its most extreme, digital-rights management technologies could control how often you listen to a song you've downloaded, where that song gets stored, whether you can share that song with someone else, and how long you have the right to listen. The technology could enable almost any form of control the copyright owner could imagine.

Copyright law supports this control in the digital age because of a deceptively simple fact about the architecture of copyright law, and the architecture of digital technology. The law regulates "reproductions" or "copies." But every time you use a creative work in a digital context, the technology is making a copy. When you "read" an electronic book, the machine is copying the text of the book from your hard drive, or from a hard drive on a network, to the memory in your computer. That "copy" triggers copyright law. When you play a CD on your computer, the recording gets copied into memory on its way to your headphones or speakers. No matter what you do, your actions trigger the law of copyright. Every action must then be justified as either licensed or "fair use."

Because every use triggers the law of copyright, I say that copyright law supports the technologies used to implement an RO culture. For if DRM says you can read an e-book only twice, all that the technology is doing is implementing a right that copyright law gives the copyright owner. Copyright law is triggered every time you read an e-book. Unless protected by "fair use," each time you read the e-book, you need permission from the copyright owner.

It's critical to recognize, however, that this control is *radically greater* than the control the law of copyright gave a copyright owner in the analog world. And this change in the scope of control came not from Congress deciding that the copyright owner needed more control. The change came instead because of a change in the platform through which we gain access to our culture. Technological changes dramatically increased, and the scope of control that the law gave copyright owners over the use of creative work increased dramatically.

In the physical world, copyright law gives the copyright owner of a book no legal control over how many times you read that book. That is because when you read a book in real space, that "reading" does not produce a copy. And because copyright law is not triggered, no one needs any permission to read the book, lend the book, sell the book, or use the book to impress his or her friends. In the physical world, the law of copyright is triggered when you take the book to a copy shop and make fifty copies for your friends—no doubt a possible use, but not an ordinary use. Ordinary uses of the book are free of the regulation of the law. Ordinary uses are *unregulated*.

But in the digital world, the same acts are differently regulated. To share a book requires permission. To read a book requires permission. To copy a paragraph to insert into a term paper requires permission. All the ordinary uses of a creative work are now regulated because all the ordinary uses trigger copyright law—because, again, any use *is* a copy.[18]

RO culture in the digital age is thus open to control in a way that was never possible in the analog age. The law regulates more. Technology can regulate more effectively. Technology can control every use. The law ratifies the control that technology would impose over every use. To the extent the copyright owner wants, and subject to "fair use," the use of his copyrighted work in digital space can be perfectly controlled. In this sense, the law supports RO culture more than it ever has.

The law's relationship to RW culture is different. For again, the very act of rewriting in a digital context produces a copy; that copy triggers the law of copyright. Once triggered, the law requires either a license or a valid claim of "fair use." Licenses are scarce; defending a claim of "fair use" is expensive. By default, RW use violates copyright law. RW culture is thus presumptively illegal.

For most of American history it was extraordinarily rare for ordinary citizens to trigger copyright law. For most of American history, the practice of RW culture would have flown under the radar of this form of commercial regulation. The reason again was a combination of the architecture of copyright law and the technologies of culture. From 1790 until 1909, copyright law didn't regulate "copies." Its core regulation was of "printing, reprinting, publishing and vending." People engaged in RW culture were not "printing, reprinting, publishing and vending." This core was expanded in 1870, when the law added the regulation of what we call derivative works to its scope. But even then, the regulation was not general. It was specific to particular activities, and these activities were again primarily commercial. "By 1873," writes Professor Tony Reese, "the subject matter of copyright protection included 'any book, map, chart, dramatic or musical composition, engraving, cut, print, or photograph or negative thereof, or...painting, drawing, chromo, statue, statuary, and...models or designs intended to be perfected as works of the fine arts.'"[19] Again, not the sort of things that ordinary RW creators do.

In 1909, the law changed. For the first time, the word "copy" was used generally to refer to the rights of any copyright owner, including the copyright owner of books. Until that time, the exclusive right to "copy" was limited to works such as statues, but not generally to works such as books. (Thus, to "copy" a statue required the permission of the copyright owner, but to "copy" a book did not.) But the revisers of the 1909 act forgot this distinction and granted copyright owners "the semblance of a right to an activity that was to have increasing importance in the new century."[20] Over time, the word began to reach every technology that "cop[ied]." Thus, as the range of technologies that enabled people to "copy" increased, so too did the effective scope of regulation increase.[21]

This automatic expansion in regulation was not really remarked upon by Congress. Congress didn't intend the expansion. Neither did it stop it. Those who benefited from this expansion were happy for the benefit. Those who might potentially be harmed by it failed to notice that the contours of their freedom—their RW freedom—were narrowing.

There were, of course, important exceptions. In the late 1960s,

Xerox gave us a technology to copy texts. In the mid-1970s, technologists gave consumers a device designed to record television shows. That device thus "copied" copyrighted content without permission of the content owner. In 1976, Disney and Universal filed a lawsuit against the maker of that device, Sony. Eight years later, the Supreme Court excused the copying, finding that consumers' actions were protected by the doctrine of "fair use."[22]

Similarly, in the 1980s there was an explosion of devices for copying cassette tapes. The content owners complained that people were using these devices to copy copyrighted content without the permission of the copyright owner. Congress ordered an extensive study of the practice. That study concluded that 40 percent of people ten and over had tape-recorded music in the past year; that the tapers had a greater interest in music than nontapers did; that most were "place-shifting"—"shifting" the "place" the copyrighted material would be played; that taping displaced some sales and inspired others; and that both tapers and nontapers believed "that it was acceptable to copy a prerecorded item for one's own use or to give to a friend."[23]

Each of these conflicts was not really a conflict with the core of RW culture. Mixtapes are perhaps an exception—many of my friends growing up thought the creativity in the selection was among the most difficult, and important, for anyone who knows anything about popular culture. But these weren't really the target of the copyright owners. Their concern instead was the use of these technologies to displace a market they thought they owned. It was technology competing with the protection of copyright. Technology, in other words, competing with RO culture. Despite this competition, the technology was largely left alone.

When RW culture moves to the Net, however, things change

dramatically. First, digital technologies, as I've already described, explode the demand for RW culture. More and more people use technology to say things, and not simply with words. Music is remixed; video mash-ups proliferate; blogs begin to build a culture around the idea of talking back.

Second, digital technologies also change how RW culture and copyright interact. Because every use of creative work technically produces a copy, every use of creative work technically triggers copyright law. And while many of these uses might be fair use, or uses licensed, expressly or implicitly, by the copyright owner, the critical point to recognize is that this is still a vast change to the history of American copyright law. For the first time, the law regulates ordinary citizens generally. For the first time, it reaches beyond the professional to control the amateur—to subject the amateur to a control by the law that the law historically reserved to professionals.

This is the most important point to recognize about the relationship between the law and RW culture. For the first time, the law reaches and regulates this culture. Not because Congress deliberated and decided that this form of creativity needed regulation, but simply because the architecture of copyright law interacted with the architecture of digital technology to produce a massive expansion in the reach of the law.

This change is also reflected further in professional culture itself. Think, for instance, about how the law regulates music. Perhaps the most distinctive American form of music is jazz. Jazz musicians create by building upon the creativity of others before. They listen to the work of others. They remake it. "Improvisation is a key element of the form." Indeed, the genre is known for its "collective improvisation."[24] Great jazz musicians are known for their ability to improvise. Louis Armstrong "essentially re-composed pop-tunes he played."[25]

A modern equivalent to jazz is called by some "laptop music," by others simply "sampling." Musicians create this music by taking the sound recordings of other musicians and remixing them. Girl Talk is an example of this. Dean Gray, Shitmat, 9th Wonder, and Doormouse are others.

The law was fairly relaxed about the creativity of jazz musicians.[26] But the law is not at all relaxed about the creativity of modern laptop music. In a series of cases beginning in the 1980s, samplers have faced an increasingly hostile judiciary which insists that any use of a recording requires the permission of the copyright owner. The point of this series came in 2004, when the Sixth Circuit Court of Appeals held that every sample used in a remixed recording triggered copyright law. There was no "de minimis" exception to copyright* that would permit samplers to avoid licensing the sample they used.[27] Beginning with hip-hop, which introduced sampling to popular culture, and continuing through laptop music today, no creative act would be distributed free of a legal cloud.

You might think that artists would be eager to end this insanity. In fact, among their lawyers at least, this craziness is a kind of lottery system. An extraordinary effort is devoted by lawyers to identifying samples used without permission in successful records. The threat of copyright liability is huge, so the payoff to make litigants go away is also huge. The system loves the game; the game thus never ends.

But this is much more than a game. There's a profound injustice

---

* Meaning simply that the amount of "copying" was so small as to not be within the scope of copyright law's concern.

in the difference of the law here, especially as it affects an emerging class of artists. Why should it be that just when technology is most encouraging of creativity, the law should be most restrictive? Why should it be effectively impossible for an artist from Harlem practicing the form of art of the age to commercialize his creativity because the costs of negotiating and clearing the rights here are so incredibly high?

The answer is: for no good reason, save inertia and the forces that like the world frozen as it is.

Can we imagine a movement that will get us from this world to a better world? Are we stuck in these dark ages? We'll think a bit more about those questions at the end of this book.

The point for now is simply to recognize that the law strongly favors RO culture while strongly disfavoring RW. Given all the good RW might do, we as a society should at least decide whether this bias against RW creativity makes sense and whether it should continue.

Lessons About Cultures

I've described two cultures of creativity. I've argued that both are important and valuable. Differences exist. In the last chapter, I described some of those differences. What can we learn from these two cultures? What lessons can we draw from how they interact?

RO Culture Is Important and Valuable

I've had lots of nice things to say about RW culture. That doesn't mean there isn't an equal amount of good to say about RO culture.

In building the Library of Congress, Jefferson hoped, as the library boasts in its slogan, "to sustain and preserve a universal collection of knowledge and creativity for future generations." That access is important because it teaches us about our past, and about the diversity of culture that lives around us. The first step of learning is listening. RO culture is essential to that first step.

RO Culture Will Flourish in the Digital Age

For most of our history, universal access was just another Jeffersonian dream. It is now a possibility. As the costs of access drop, there will be a market incentive to build the biggest "library" in human history. And like the very best libraries in our past, the job of this library will be to assure access. Not necessarily for free, for in many cases, "free" would produce insufficient incentives to build that access. Instead, RO culture will flourish; more culture will be accessible more cheaply than at any point in human history.

RW Culture Is Also Important and Valuable

Yet the history of the Enlightenment is not just the history of teaching kids to read. It is also the history of teaching kids to write. It is the history of literacy—the capacity to understand, which comes not just from passively listening, but also from writing. From the very beginning of human culture, we have taught our kids RW creativity. We have taught them, that is, how to build upon the culture around us by making reference to that culture or criticizing it. As

Negativland's Hosler put it to me: "*Of course* human beings build on what came before them in anything they create. That's just obvious." We have encouraged them to build upon it. We have forced them to acquire a "literacy" about the culture around them. We test our kids on the basis of that literacy; we reward the "literate" of every generation. We reward "writing."

For most of human history, text was the only democratic literacy. For most of human history, words were the only form of expression that everyone had access to. The twentieth century gave us an extraordinary range of new types of "writing." But until the last years of that century, none of that "writing" was ever democratized in the way that text had been. Only a few could go to film school. Only a relatively few had the resources to learn how to record or edit. The single most important effect of the "digital revolution" was that it exploded these historical barriers to teaching. Every important form of writing has now been democratized. Practically anyone can learn to write in a wide range of forms. The challenge now is to enable this learning, not only by building the technologies it requires, but by assuring the freedom that it requires.

So again, as I encouraged in chapter 2, think a bit about that freedom. Remember when you learned to write. Remember the act of quoting. Or incorporating. Or referring. Or criticizing. What freedoms did you take for granted when you did all of this? Did you ask permission to quote? Did you notify the target of your criticism that you were criticizing him? Did you think twice about your right to dis a movie you saw in a letter to a friend? Were you ever troubled by quoting Bob Dylan in an essay about war?

The answer to all these questions is of course "no." We grew up taking for granted the freedoms we needed to practice our form of writing. We created, and we shared our creativity with whoever would read it (our parents and teachers, if we were lucky). We never questioned the right to create in this way, freely.

Our kids want the same freedom for their forms of writing. For not just words, but for images, film, and music. The technologies we give our kids give them a capacity to create that we never had. We've given them a world beyond words. This world is part of what I've called RW culture. It is continuous with what has always been part of RW culture—the literacy of text. But it is more. It is the ability for amateurs to create in contexts that before only professionals ever knew.

WHETHER RW CULTURE FLOURISHES DEPENDS AT LEAST IN PART UPON THE LAW

As it exists now, copyright law inhibits these new forms of literacy. I don't mean that it stops kids from remixing. No law could ever do that, any more than a law could stop quoting. Instead, I mean that the law as it stands now will stanch the development of the institutions of literacy that are required if this literacy is to spread. Schools will shy away, since this remix is presumptively illegal. Businesses will be shy, since rights holders are still eager to use the law to threaten new uses. Uncertainty about the freedom to engage in this form of creativity will only stifle the willingness of institutions to help this form of literacy develop. RW culture can't help but expand the sense of "writing." But legal culture will force the institutions that teach writing to stay far away from this new expressive form.

THE LAW'S CURRENT ATTITUDE IS BOTH DESTRUCTIVE AND SELF-DEFEATING—TO VALUES FAR MORE IMPORTANT THAN THE PROFITS OF THE CULTURE INDUSTRIES

We, as a society, can't kill this new form of creativity. We can only criminalize it. We can't stop our kids from using the technologies we give them to remix the culture around them. We can only drive that remix underground. We can't make our kids passive in the way we were toward the culture around us. We can only make them "pirates." So does this criminalization make sense?

Here history has an important lesson. About a decade ago, scholars and activists began calling for a legislative response to what we would eventually term "peer-to-peer file sharing." We did not call for further penalties for illegal file sharing. Instead, we called for decriminalization. A wide range of proposals asked Congress to create a compulsory license for peer-to-peer (p2p) file sharing. Under this license, the act of sharing music, for example, would not violate any law of copyright. It would instead simply affect how compensation for file sharing would be shared.

The most ambitious of these proposals was made by Professor William Fisher at Harvard. Under Fisher's plan, Congress would permit, for example, music to be shared freely.[28] It would establish technologies to sample who was getting what. Then, based upon that evidence of popularity, it would compensate artists for their creativity through a tax that would be imposed upon digital technologies in the most efficient way possible. Thus, Madonna would make more than Lyle Lovett; and Lyle Lovett would make more than I.

There are plenty of ways to criticize these proposals. While I've made my own, I've also criticized many others. But as we look back over the last ten years and imagine how things would have been different if, in 1998, Congress had enacted any of these proposals, several facts become clear.

First, the war on file sharing has been an utter failure. As one article notes,

> More than 5 billion songs were swapped on peer-to-peer sites [in 2006] while CD sales, the industry's core revenue-producing product, continue to decline, dropping about 20 percent this year alone. And according to a recent report from Jupiter Research, things are only going to get worse. "Young consumers are increasingly shunning music buying in favor of file-sharing, which is four times more popular than digital-music buying among ages 15 to 24," the report notes.[29]

A picture captures the point much better than words. Using data provided (generously for free) by BigChampagne Online Media Measurement, we can graph the average number of p2p users from August 2002 to October 2006 (see pages 112–13). The gray bar indicates when the Supreme Court decided *MGM v. Grokster,* holding p2p file sharing to be illegal.[30] As is plain, whatever the law is doing, it's not having much effect upon what p2p users are doing.

Second, had a compulsory license been put in place, artists would have received more money over the last ten years than they have. Legal sharing may have stanched some growth in legal sales. But there was a huge amount of content shared "illegally" that, under this alternative, would at least have triggered compensation for artists.

Third, had businesses been free to rely upon these licenses, there would have been an explosion in innovation around these technologies. It would not just have been a few who could strike deals with the ever terrified recording industry. Anyone who had an idea could have deployed it, consistent with the terms of the compulsory license. Thus, innovation in content distribution would have been greater too.

But fourth and most important, had we had a system of compulsory licenses a decade ago, we wouldn't have a generation of kids who grew up violating the law. As a recent survey by the market research firm NPD Group indicated, "more than two-thirds of all the music [college students] acquired was obtained illegally."[31] Had the law been changed, when they shared content, their behavior would have been legal. Their behavior would therefore not have been condemned. They would not have understood themselves to be "pirates." Instead, they would have been allowed to lead the sort of childhood that I did—where what "normal kids did" was not a crime.

Again, I don't mean this as an argument in favor of decriminalizing all currently illegal behavior. Whether or not kids rob banks, it should be illegal to rob banks. The wrong of rape is increased, not mitigated, by its frequency. Instead, I'm asking you to weigh one bad against the other: What our policy makers have done over the last decade has not actually stopped file sharing; it has not actually helped a lot of artists; it has not spurred a wide range of innovation. All it has done with certainty is raise a generation of "pirates."

Weigh that bad against the alternative: if there had been a compulsory license, artists would have had more money; business (outside of the recording industry) would have had a greater opportunity to innovate; and our kids would not have been "pirates."

Average Simultaneous P2P Users—USA

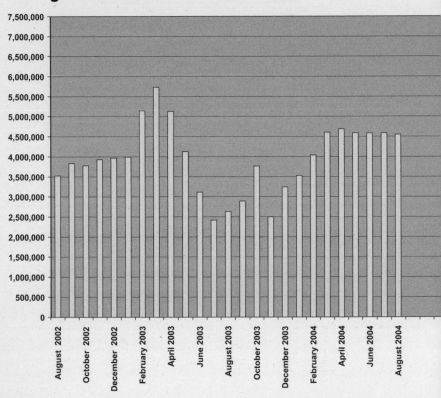

No doubt someone would have lost something in this alternative scenario. Lawyers for sure. Maybe record companies too. Lawyers would have missed out on the extraordinary boon to our industry caused by the litigation surrounding illegal file sharing. Record companies might have lost out because they would have given up an exclusive right in favor of compulsory compensation. But even this loss is uncertain. It is more than plausible that a compulsory system would have secured for the recording industry more money than in fact it got.

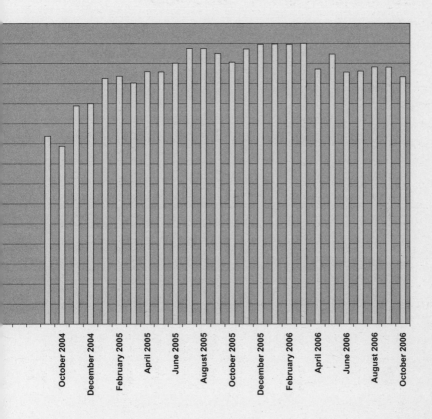

I raise this parallel not to endorse peer-to-peer file sharing. I stand by my position in *Free Culture* that "piracy" is wrong— a position I repeated nine times in that book.[32] Instead, I raise the parallel to ask whether we want to make this mistake again. Should the next ten years be another decade-long war against our kids? Should we spend more of our resources hiring lawyers and technologists to build better weapons to wage war against those practicing RW culture? Have we learned nothing from the total failure of policy that has defined copyright policy over the last decade?

I believe this for the same reason the content industry is so keen to enforce copyright. As the RIAA's Mitch Bainwol and Cary Sherman explained:

> It's not just the loss of current sales that concerns us, but the habits formed in college that will stay with these students for a lifetime. This is a teachable moment—an opportunity to educate these particular students about the importance of music in their lives and the importance of respecting and valuing music as intellectual property.[33]

Exactly right. So what rules should we work so hard to enforce?

The argument in favor of reforming our legal attitude toward remixing is a thousand times stronger than in the context of p2p file sharing: this is a matter of literacy. We should encourage the spread of literacy here, at least so long as it doesn't stifle other forms of creativity. There is no plausible argument that allowing kids to remix music is going to hurt anyone. Until someone can show that it will, the law should simply get out of the way. We need to decriminalize creativity before we further criminalize a generation of our kids.

PART TWO

ECONOMIES

In the first part of this book, I described two cultures: one made to be consumed (RO), the other made to be remade (RW). Both cultures were part of our past: RW creativity from the dawn of human culture, RO from the birth of technology to capture and spread tokens of culture. Both, I've argued, will be part of our future: The Internet will enable a more vibrant RO culture. It will also enable a more expansive RW culture.

But to see just how both futures will unfold, we need to think a bit less about culture, more about economics. Nothing gets built unless there are the incentives to build it. And while the incentives to build RO culture are clear enough, what builds RW culture? Who will invest in its construction? Who's going to pay for customer support?

In this part, I describe two economies of social production— commercial and sharing economies—on the way to describing a mix between these two, an economy I call a "hybrid." In the sense that I will describe, we have seen hybrids before. They have been part of our history. But their significance is now, potentially at least, radically increased. Every interesting Internet business will be a hybrid. Understanding how, and how to do it well, will become the most important challenge for Internet entrepreneurs.

SIX

TWO ECONOMIES: COMMERCIAL AND SHARING

An "economy" is a practice of exchange that sustains itself, or is sustained, through time. A "practice of exchange." For example:

1. A gives something to B.
2. B (directly or indirectly) gives something back to A.
3. Repeat.

The "something" could have tangible, economic value—money, or hours of labor. Or it could be intangible, and without ordinary economic value—friendship, or helping a neighbor with a flat tire. In either case, the trade occurs within an "economy" when it is a regular practice of social interaction. People participate within that economy so long as they get enough back relative to what they give. This doesn't mean everyone gets back exactly what he or she contributes (or more): A talented lawyer working for a public-interest-housing law firm gives more than her meager salary returns. (Meet my wife.) But it does mean that people operating within an economy evaluate the exchange, how much they get versus how much they give, and that we should expect they will continue in that economy so long as they get enough from the exchange relative to what they give.

"Economies" in this sense differ in many ways. In the story that follows, however, I radically and crudely simplify these differences to speak about three types of economies only: a commercial economy, a sharing economy, and a hybrid of the two.

Following the work of many, but in particular of Harvard professor Yochai Benkler,[1] by a "commercial economy," I mean an economy in which money or "price" is a central term of the ordinary, or normal, exchange. In this sense, your local record store is part of a commercial economy. You enter and find the latest Lyle Lovett CD. You buy it in exchange for $18. The exchange is defined in terms of the price. This does not mean price is the only term, or even the most important term. But it does mean that there is nothing peculiar about price being a term. There's nothing inappropriate about insisting upon that cash, or making access to the product available only in return for cash.

A "sharing economy" is different. Of all the possible terms of exchange within a sharing economy, the single term that isn't appropriate is money. You can demand that a friend spend more time with you, and the relationship is still a friendship. If you demand that he pay you for the time you spend with him, the relationship is no longer a friendship.

So, again, there's nothing odd about your local Wal-Mart insisting that you give them $2.50 for a bottle of juice. You might not like that demand; you might well think $2.00 is the right price. But there's nothing inappropriate about Wal-Mart's demand. In our culture at least, that's just the way a Wal-Mart is supposed to deal with us.

Nor is there anything odd about a softball team demanding that every member make at least ten of the twelve games in a season. Again, you might not like the demand; you might wish you could miss four rather than just two games. But there's nothing inappropriate about the team's demand. Indeed, it is a perfectly reasonable way to make sure participants within this sharing economy actually participate.

But it would be very odd if a friend apologized for missing lunch and offered you $50 to make it up. And it would be very, very odd if your girlfriend, at the end of a great date, offered you $500 to spend the night. Or if Wal-Mart asked all customers to "pitch in and help Wal-Mart by sweeping at least one aisle each time you shop." Or if McDonald's asked you to "help out" by promising to buy hamburgers at least once a month. Money in the sharing economy is not just inappropriate; it is poisonous. And "helping out" is not just rare in a commercial economy. It is downright weird.

Viewed like this, we all live in many different commercial and sharing economies, all at the same time. These economies complement one another, and our lives are richer because of this diversity. No society could survive with just one or the other. No society should try.

The Internet has many examples of commercial and sharing economies. In this chapter, we consider examples of both. But the aim throughout this chapter is simply preliminary: to set up a richer understanding of a much more interesting phenomenon—the hybrid—that we'll consider in the chapter that follows.

Commercial Economies

My wife (to be) and I were at an Italian restaurant. In Italy. I ordered pasta with a mushroom sauce. She ordered pasta with a tomato sauce. After the waiter served us, he offered my wife some Parmesan cheese. She accepted. He grated the cheese for her and then began to leave. I stopped him and said I too wanted Parmesan cheese.

"No," he told me.

"No?" I asked.

"No," he said again. "The cheese would overwhelm the taste of the mushrooms."

Startled a bit, I hesitated. "But what if I want the taste overwhelmed?"

"That's not my concern," he informed me, and walked away.

You learn a great deal about who you are by noticing the things that enrage you, and then working out just why. This was one of those moments for me. What "right" did this waiter have to interfere with my eating my pasta as I wanted? It wasn't as if I was going to complain to the locals about the taste of the pasta. Nor was I likely ever to return to this village or this restaurant. Indeed, to the contrary, the exchange made me resolve never to return to this restaurant. The waiter was out of line. I would take my business elsewhere.

My reaction came from a certain view I held (unnoticed until that moment) of my relationship to a restaurant. It was, in the sense I mean in this chapter, for me simply a "commercial" relationship. This was a transaction within a "commercial economy." Had the waiter said, "Sure, extra cheese costs an extra euro (for both you and your wife)," that would have been totally appropriate. Price is how we, in commercial economies, negotiate things. If there's something I want that they want to ration, then let them use the market's most ubiquitous tool for rationing—money. The business of business is to make money—not, as this waiter saw it, to avoid insulting mushroom-strewn pasta.

Of course, there's nothing natural or necessarily right about my view of my relationship to this beautiful Italian restaurant. But I take it that anyone living within a modern commercial economy would have the same or a similar response in at least some part of the commercial economy. Imagine the dry cleaner who refused to clean an old sweater: "That design went out ages ago." Or a coffee shop that insisted, "Tell us about your day!" before the barista would take your order. ("It's the friendly way to be!" the shop insists.) All of us, somewhere in our life, relish the simplicity of the market. And some of us (myself included) yearn for ways to make more of our life governed by the simple logic of markets.

If we're in a place where we feel such simplicity should reign, where we're not insulted when someone mentions money, where we meter the relationship with price, then we're within a "commercial economy." The market is the engine that drives this commercial economy; if well designed (meaning regulated to protect participants from force or fraud), the market is an extraordinary technology for producing and spreading wealth. The commercial economy is a central part of modern life; it has contributed to human well-being perhaps more than any other institution created by humankind. We are well beyond the point where it makes sense to oppose the flourishing of the market.

A critical part of the Internet is just such a commercial economy. Indeed, for some, it is the most important part. The Internet has caused an explosion in the opportunities for business to make money by making old businesses work better. It has also made possible new businesses that before the Net weren't even conceivable. And while we're just beginning to get a clear sense of what makes business prosper on the Internet, we can already see that this new bit of social infrastructure offers a staggering potential for growth and innovation. In 1994, there were 1,700 dot-com domain names. Twelve years later, there were more than 30 million.[2] There was no category called the "e-commerce sector" in 1994. In 2005, the e-commerce sector was estimated to be worth $2.4 trillion: $1.266 billion for manufacturing, $945 billion for wholesale, $93 billion for retail, and $96 billion for selected service industries.[3]

What makes the Internet's commercial economy work? Or why does it work so well, or differently from real-space economies? In the balance of this section, we will consider a few key features that explain its success. My aim is saliance, not comprehensiveness. I want only to draw out a few features that will make the relationships among commercial and sharing economies clear.

Three Successes from the Internet's Commercial Economy

Begin with some familiar examples of Internet success, from which we can draw some lessons of success.

NETFLIX

More than thirty years ago, in November 1976, America's film industry launched a war against a technology that was quickly becoming ubiquitous: the (what we now call) VCR. The VCR had been designed to record programming "off air." Most of that programming was copyrighted. Copying copyrighted works without the permission of the copyright holder was, Universal and Disney claimed, a crime. The VCR, they thus argued, was a tool designed to enable a crime.

Eight years later, the Supreme Court disagreed.[4] By a close vote, the Court held that the VCR itself was not illegal, because although it could be used to infringe, it was also "capable of a substantial noninfringing use." At least some of the copyright owners, the Court noted, whose work would be taped were happy that it would be taped. (Mr. Rogers was the Court's favorite example.) And for those not happy that their work was being recorded, the Court held that at least sometimes, this "time-shifting" of content was a protected "fair use." These noninfringing uses thus saved the technology from being banned by copyright law. Hollywood would have to figure out how to make money despite the technology.

In the thirty years since Hollywood lost that case, it has become clear just how lucky it was to lose. Video sale and rental revenues far surpass what the film industry makes in the theater.[5] Had the studios won, it's not clear just how much the platform of that success would have spread.

Blockbuster Video was a key reason losing the war on VCRs was a victory for Hollywood. For the Blockbusters of the world soon brought more revenue to Hollywood than its own blockbusters in theaters did. The store first launched in Dallas in 1985, with eight thousand tapes and 6,500 titles. Because of the spread of VCRs, there was a ready infrastructure to support Blockbuster's business. Two years later there were fifteen stores and twenty franchises. By mid-1989 there were more than seven hundred stores. At the end of that year there were one thousand stores.[6]

Blockbuster was a brilliant innovation in the distribution of film. But it had important drawbacks. However convenient it made finding a film, you still had to go to the Blockbuster and browse through endless fluorescent-lit aisles of videos to find it. And however endless those aisles of videos, each Blockbuster could in fact carry a relatively small number of films. There was thus more choice than TV, and on your own schedule. But not endless choice. And though convenient, the system still had its costs.

In 1997, Reed Hastings had a better idea for delivering video to consumers. Rather than a harshly lit store at a strip mall, Hastings thought, the Internet would be a pretty good way to browse for films. Indeed, using smart preference-matching technologies, the Internet would be a better way to browse for film because the machine would help you find what in fact you wanted. He thus launched one of the Internet's most famous success stories: Netflix. Customers paid Netflix a flat monthly fee; in exchange, they could rent DVDs of favorite films; those DVDs were sent through the mail, with simple return envelopes included; the monthly subscription entitled the customer to hold a fixed number of DVDs. Thus, if you had a three-DVD subscription, you paid about $17 a month. You ordered three movies that you wanted to see, and Netflix sent them. You could hold on to these movies for as long as you wanted (hence, no late fees). And when you returned one, the next on your queue was sent. The only inconvenience of this system was that you had to plan ahead a bit. The great advantage was that if you planned a bit in advance, the films would be waiting at home whenever you wanted to watch them.

Netflix has radically changed the video rental market. The best evidence of its effect is that in 2004, Blockbuster changed its business model to mirror Netflix's to better compete.[7] Wal-Mart's service was taken over by Netflix in 2005.[8] Hastings's model thus became the industry standard.

AMAZON

Perhaps the first dramatically successful example of Internet commerce began as a simple bookstore. Founded in 1994 (as Cadabra .com) and launched in 1995 (as Amazon.com), Amazon set out to do more efficiently what bookstores had always sought to do: sell books. When the online store opened, it had only two thousand titles in stock. But within the first month it had orders from all fifty states and from forty-five countries outside the United States. Sales in 1997 reached approximately $150 million. Two years later, sales were $1.6 billion. Two years after that, because of third-party deals with companies such as Target and America Online, sales exceeded $3 billion. In 2003 the company crossed $5 billion. In 2006 sales totaled more than $10 billion.[9]

Once again, this store had advantages very similar to the advantages of Netflix. Rather than browsing a Barnes & Noble superstore, the customer used his computer to see what books there were to buy. And rather than the customer using his car to collect the books he wanted, Amazon used the U.S. Postal Service. Amazon founder Jeff Bezos's bet was that the convenience of browsing would outweigh the delay in receipt. More important, Amazon could far surpass any bricks-and-mortar store in the size of its inventory.

Amazon's success, however, didn't come naturally. The company has been relentless in building innovation to drive sales. In 2003 the company launched the Amazon Associates program, which enabled independent sites to become sellers for Amazon. The Associates earn revenue from referrals to Amazon. In 2005 it launched Amazon Connect, which enabled authors to post remarks on the book pages for their books. In 2006 the company launched Amazon S3, offering high-bandwidth storage and distribution for large digital objects. In January 2007 the company began Ama-pedia, "a collaborative wiki for user-generated content related to 'the products you like the most.'"[10] In the decade since Amazon launched, it has delivered to the market an extraordinary range of innovation. Everything it does is aimed to drive sales of its products more efficiently.

One of the techniques that Amazon uses mirrors the technique of the Internet generally: Amazon has opened its platform to allow others to innovate in new ways to build value out of Amazon's data-base. Through a suite of tools called Amazon Web Services (AWS), Amazon enables developers to build products that integrate directly into Amazon's database. For example, a developer named Jim Bian-colo used AWS to build a free Web tool to track the price difference between new products and used products (plus shipping). And a company called TouchGraph used AWS to build a product browser that would show the links between related products. Enter Cass Sunstein's, for example, and you'll see all the books in Amazon that relate to Sunstein's books in subject and citation.

Amazon sells some of these AWS services. Some it leaves free. But it develops these services if it believes such development will drive the sales of its products, and perhaps even teach Ama-zon something about how to better offer its products. Of course, it ultimately controls the platform. What others add, Amazon can take away. But in a limited way, the platform invites innova-tion from others. That innovation rewards others and Amazon both.

GOOGLE

Without a doubt, the most famous example of Internet success is Google. Founded at Stanford by two students (the first URL was http://google.stanford.edu), the company radically improved the effectiveness of Internet searches. Rather than selling placement (which can often corrupt the results) or relying upon humans to index (which would be impossible given the vast scale of the Internet), the first Google algorithms ordered search results based upon how the Net linked to the results—a process called PageRank, referring not to "page" as in Web page, but "Page" as in Larry Page, Google cofounder and developer of the technique.[11] If many Web sites linked to a particular site, that site would be ranked higher in the returned list than another Web site that had few links. Google thus built upon the knowledge the Web revealed to deliver back to the Web a product of extraordinary value. The company was founded in 1998. In 2005 its market capitalization was $113 billion; in July 2007 it had risen to $169 billion.[12]

One might well say that all of Google's value gets built upon other people's creativity. Google's index is built by searching and indexing content others have made available on the Web. As I've described, the original algorithm built its recommendations upon the links it found already existing on the Web; later, the algorithm also adjusted its recommendations based upon how people responded to the results Google returned. In all of these cases, the value Google creates comes from the value others have already created.

Some draw a downright foolish conclusion from the fact that Google's value gets built upon other people's content. Andrew Keen, for example, a favorite from chapter 5, writes, "Google is a parasite; it creates no content of its own."[13]

But in the same sense you could say that all of the value in the *Mona Lisa* comes from the paint, that Leonardo da Vinci was just a "parasite" upon the hard work of the paint makers. That statement is true in the sense that but for the paint, there would be no *Mona Lisa*. But it is false if it suggests that da Vinci wasn't responsible for the great value the *Mona Lisa* is.

Like Amazon, Google also offers its tools as a platform for others to build upon. We'll see this more below as we consider Google Application Programming Interfaces (APIs). And more successfully than anyone, Google has built an advertising business into the heart of technology. Web pages can be served with very smartly selected ads; users can buy searches in Google to promote their own products.

The complete range of Google products is vast. But one feature of all of them is central to the argument I want to make here. Practically everything Google offers helps Google build an extraordinary database of knowledge about what people want, and how those wants relate to the Web. Every click you make in the Google universe adds to that database. With each click, Google gets smarter.

Three Keys to These Three Successes

These familiar stories of Internet success reveal three keys to success in this digital economy.

LONG TAILS

The first of these three is also perhaps the most famous. Each of these three Internet successes takes advantage of a principle that

Amazon's Jeff Bezos recognized in 1995, and that *Wired*'s editor in chief, Chris Anderson, formalized in 2005 in his book *The Long Tail*.[14]

The Long Tail principle (LTP) says that as the cost of inventory falls, the efficient range of inventory rises. And as transaction costs generally fall to zero, the efficient inventory rises to infinity. Put differently, the less it costs to hold a particular book or DVD in inventory, the more books or DVDs a particular company can profitably hold. Thus, Amazon can offer its customers more books than any bricks-and-mortar store could, since it can store these books efficiently at inventory locations around the country. And more important, a big share of Amazon's profits come from titles that are unavailable anywhere else. Chris Anderson estimated that 25 percent of Amazon's sales come from its tail (where the tail represents products not available in a bricks-and-mortar store). More generally, the current data at Rhapsody, Netflix, and Amazon show that the tail amounts to between 21% and 40% of the market.[15] Netflix profits in the same way. Netflix offers seventy-five thousand titles today (about twelve thousand in 2002) in more than two hundred genres on its Web site. Blockbuster offered seven thousand to eight thousand in 2002.[16]

The Long Tail dynamic benefits those whose work lives in the niche. A wider diversity of films and books is available now than ever before in the history of culture. The low cost of inventory means wider choice. Wider choice is a great benefit for those whose tastes are different.[17]

Those who doubt the significance of the Long Tail are quick to argue that the amount of commerce generated in the Long Tail is small relative to the market generally. Anderson calculates 25 percent of Amazon's sales come from its tail; but the *Wall Street*

Journal's Lee Gomes comments, "[U]sing another analysis of those numbers...you can show that 2.7% of Amazon's titles produce a whopping 75% of its revenues."[18]

But this criticism misses two important points. First, all the excitement in a market is action at the margin. Like with runners in a 100-meter dash, the difference between first and last place may be just .02 seconds. But that is the difference that matters, and the difference produced by sales in the Long Tail will matter lots to companies struggling to compete.

Second, and more important, the breadth of this market will support a diversity of creativity that can't help but inspire a wider range of creators. For reasons at the core of this book, inspiring more creativity is more important than whether you or I like the creativity we've inspired.

Perhaps the best evidence of this comes from another increasingly successful example of this Internet economy, launched by one of the key entrepreneurs changing an operating system called Linux from a hobby to a business: Red Hat and its cofounder Robert Young. After Red Hat went public in 1999, Young moved on to start his next great idea: Lulu Inc., a technology company that helps people "publish and sell any kind of digital content."

Lulu's aim is to out-Amazon Amazon, to "put all the books in a bookstore that can't fit on Amazon."[19] The market is not the niche that Amazon's Long Tail serves, but the "small niche market" that is beyond even Amazon's reach. As Young told me, "Amazon's business model is built around the business model of the existing book-publishing industry. Lulu's business model is a completely different Internet-based business model that...doesn't even look at what the publishing industry does."

Lulu does this by working hard to educate authors about how

best to write to compete. "If you're going to write a detective novel," Young explained to me, "that competes with Agatha Christie, figure out what your hook is." "Why should your detective novel sell?" Lulu asks its authors. "Is there something unique about it?"

Lulu's aim is not to spread free culture, if that means culture you don't have to pay for.[20] "We think sharing is easy," Young told me. "What's difficult is empowering people to actually get paid for content they are producing." Lulu focuses not on all of the "ninety-nine out of a hundred" authors who get rejected by the traditional publishing market. Instead, it focuses on the "forty-nine out of a hundred": people who "actually have something valuable to say and should have a market." These are people who are

> writing for too small a market or they're writing another book on a subject that the publishers have already published a book on. Either way, the publisher doesn't want it because he doesn't see any profit in it. Not…because it's a bad book. He admits it's a valuable book. It's just he doesn't want it because he's already got two other books on [for example] programming in Java. So he doesn't want a third.

Once again, on the margin, what will make Lulu successful where vanity presses were not is the efficiency with which creative work can be produced and distributed way down the Long Tail. Young is fanatical about the challenge in selling down the tail. There's nothing automatic. It takes hard work by both Lulu and the author. Success gets made; no "Long Tail magic" makes it for anyone.

But the consequence of his success will be a much wider range of people creating. And this is the most important consequence for society generally. Just as Jefferson romanticized the yeoman farmer working a small plot of land in an economy disciplined by hard work and careful planning, just as Sousa romanticized the amateur musician, I mean to romanticize the yeoman creator. In each case, the skeptic could argue that the product is better produced elsewhere—that large farms are more efficient, or that filters on publishing mean published works are better. But in each case, the skeptic misses something critically important: how the discipline of the yeoman's life changes him or her as a citizen. The Long Tail enables a wider range of people to speak. Whatever they say, that's a very good thing. Speaking teaches the speaker even if it just makes noise.

Little Brother

The Long Tail alone, however, is not enough to explain the great success of the Amazons/Netflixes/Googles of the world. It's not enough that stuff is simply available. There must also be an efficient way to match customers to the stuff in the Long Tail. I may well want to buy a book that only five hundred others in the world would want to buy. But I'm not about to sift through the 10 million other books on Amazon's shelf to find that one that I'd be eager to buy. Amazon (and Netflix and Google) have got to do that for me. And each of these companies does it well by, in a phrase, spying on my every move. An efficient Little Brother (a relative of Orwell's Big Brother) learns what I'm likely to want and then recommends new things to me based upon what he has learned.

Collecting data about customers is, of course, nothing new. But the key to the efficiency of this Little Brother is that it builds upon a principle described best by VisiCalc co-inventor Dan Bricklin in an essay called "The Cornucopia of the Commons."[21]

Bricklin's essay was inspired by a quibble he had with those who said Napster was so successful because it was a peer-to-peer technology. Napster's success, he argued, had nothing to do with peer-to-peer. First, the system was not in fact a "peer-to-peer" technology. Second, not using a p2p architecture may well have been a better technical strategy to serving the ends that Napster sought.

Bricklin argued that Napster's success came not from a technical design, but from an architecture that produced value *as a by-product* of people getting what they wanted. When you installed Napster, by default it made shareable the music you had on your computer. The more people who joined, the better the "database." And as a Napster user added content to his library by, for example, ripping a CD, "creating the copy in the shared music directory c[ould] be a natural by-product of [his] normal working with the songs."[22] "Increasing the value of the database by adding more information is a natural by-product of using the tool for your own benefit." "No altruistic sharing motives need be present" to explain the network's extraordinary success.

Bricklin made the same point about a service called CD Database (CDDB). CDDB was originally created by volunteers who wanted a simple way to get track information about their music. CDs ship with the track identified simply by a number and a total track time. But by using cryptographic signing technologies, it's fairly easy to get a unique signature for every song on any CD. Using that signature, an Internet database can easily identify which song is on your CD if that song's signature has already been entered onto the database along with information about the song's name, artist, etc. Thus, by getting people to add that information into the database, the database becomes more valuable to everyone.

Notice a corollary to Bricklin's design law suggested by a

commenter on Bricklin's original essay, Evan Williams: Design
the database so people use the data they enter, thus increasing their
incentive to get it right.[23] Apple's iTunes does that right now. If
you put a CD into your iTunes-enabled computer, chances are it
launches iTunes. And if iTunes is connected to the Internet, iTunes
then compares the track information from the CD with the (now)
Gracenote CD database. If it finds the CD, then it substitutes the
uninteresting "Track 01, Track 02" titles provided by the CD itself
with the artist and track information. But if it doesn't find the track
information, then it informs you, and invites you to enter the data
yourself.

Once you've entered the data, iTunes then gives you a simple way
to send that data to Gracenote. Gracenote gets to choose whether
to accept the submission or not, but the point is, Gracenote knows
(because it is filtering the input through services like this) that it's
likely the data you've entered is valid. It's a hassle to enter the data
in the first place; it would be a real hassle to enter false data, submit
it, and then enter the real data. And no doubt, Gracenote can hold
inputs till it gets corroboration.

The critical point again is that the design of Gracenote elicits
the valuable data, not any particular love for Gracenote or Apple.
The design "add[s]...value [to] the database without [adding] any
extra work [to the user.]"[24]

Perhaps the best example of this kind of by-product value creation (in theory at least; the lawyers never allowed this system to get
going) was the aspiration of the company sued into the Dark Ages,
MP3.com. Michael Robertson, the company's founder, wanted to
remake the world of music production by finding a better way to
market new bands to existing customers. A strong believer in the

efficacy of Little Brother, Robertson thought the best way to market is to understand your customers perfectly. And one way to understand your customers perfectly (or as perfectly as humans can) is to see what stuff they already own.

Robertson had a brilliant, Cornucopia of the Commons way to learn just this. He gave the customers something they wanted in exchange for them giving him something he needed.

The service he gave them was called my.MP3.com. It promised to give customers access to "their music" wherever they were. To do this, customers would simply need to show MP3.com what "their music" was. The customer would submit a CD that she (presumptively) owned to a program called Beam-it. Beam-it would identify the CD and report its identity to MP3.com. MP3.com would then give the user access to that music wherever she was (on the Net at least). Thus, in exchange for learning what music customers had, MP3.com gave those customers access to their music everywhere. And then, using the complex of preference data MP3.com would collect, the company could predict which of its own catalog its customers were likely to love. So if it saw that I liked Lyle Lovett, and then saw that I liked one of its new artists too, then it would have a good reason to try to promote that new artist to others who liked Lyle Lovett. (Of course, the real algorithm was much more complex than this; but that's the basic idea.)

Once again, this design would work because it asked nothing more of its customers than the ordinary effort the customers would expend to get what they wanted. It would thus efficiently gather the data necessary to make the business work. And this ability to gather this data efficiently is a key reason Internet businesses can beat their bricks-and-mortar equivalents. Just think of the revolt there would be if Barnes & Noble superstores had clerks following you around, recording what books you looked at and which you bought. Yet this is precisely what Amazon can do, simply by designing its system well.

All three of my examples of Internet successes build upon the Bricklin insight to feed Little Brother, none perhaps as comprehensively as Google. Every Google product is designed to give a user what he or she wants and, at the same time, to gather data that Google needs. You don't have a choice about helping Google when you use Google's search engine. Your search is a gift to the company as well as something valuable to you. The company efficiently serves you a product, and very efficiently learns something in the process.

There are many who are troubled by Little Brother. Professor Jeff Rosen once described the terror and outrage he felt at knowing Amazon was "watching" what books he bought in order to recommend new books to him. When I heard his description, I realized that one of us was from a different planet. No doubt Amazon might abuse the data it collects. But also, no doubt, it has a huge incentive not to. (Unlike the U.S. government, if Amazon screws up, I can take my business elsewhere.) Anyway, it's not as if Jeff Bezos is reading my (almost daily) orders. Some computer somewhere is simply responding to input collected from me. And while I might care lots about what my neighbors, or students, or friends think about me, I don't care a whit about what some computer thinks about my tastes.

This is not to say we shouldn't be concerned with how these data might be used. When the United States government demanded that Google provide it with search queries relating to pornography in the context of the government's defense of the Child Online Protection Act, Google fought the demand fiercely in court, no doubt in part because it didn't want its users to think that their every search might be made available to the government.[25] Likewise, the company has recently taken steps to partially anonymize the data it holds, to avoid demands like this in the future and to respond to harsh criticism by privacy groups that claim Google's database is in effect a privacy time bomb.

These are important concerns, but beyond my focus here. They emphasize, however, a central design feature of the successful Internet economy: build the technology to feed Little Brother with the mouse droppings of happy customers. (Okay, that sounds gross, but you get the point.)

LEGO-ized Innovation

The final feature of these three Internet successes that I want to highlight is ultimately one that generalizes to the Internet itself. All three of these successful Internet businesses build their value in part by allowing others to innovate upon their platform. Functionality gets LEGO-ized: it gets turned into a block that others can add to their own Web site or their own business.

Netflix does this the least among the three, but it does it nonetheless. (The company was scolded by one of the Net's leading bloggers in 2004 for failing to offer APIs.[26] It is slowly responding.) Its purpose is to "improve the accuracy of predictions about how much someone is going to love a movie based on their movie preferences."[27] To achieve this end, Netflix runs a "Netflix Prize"—offering a grand prize of $1 million to anyone who improves Netflix's own system by more than 10 percent. To enable this competition to happen, Netflix shared "a lot of anonymous rating data." The company also increasingly offers through RSS feeds access to ranking information about its users' choices.

Amazon does this through its Amazon Web Services. And Google does this perhaps most of all, through Google APIs that encourage what has come to be known as the Google mash-up. Don Tapscott and Anthony Williams describe one example of the Google mash-up in their book, *Wikinomics*.

> In May 2005, Paul Rademacher was trying to find a house in Silicon Valley for his job at Dreamworks Animation. He grew weary of the piles of Google maps for each and every house he wanted to see, so he created a new Web site that cleverly combines listings from the online classified-ad service craigslist with Google's mapping service. Choose a city and a price range, and up pops a map with pushpins showing the location and description of each rental. He called his creation housingmaps.
>
> While a useful tool for helping people find a place to live, on the surface it hardly seems groundbreaking. And yet, Paul Rademacher's site quickly became a poster child for what the new Web is becoming, not because of what it was, but for how it was created. Housingmaps was one of the Web's first mashups.
>
> Google Map mashups, for example, have emerged to do everything from pinpointing the locations of particular crime sites, to outing celebrity homesteads, to enabling fitness buffs to measure their daily running distance. Or, for the price conscious, there's CheapGas, a service that mashes Google Maps and GasBuddy together to identify gas stations with the lowest pump prices.[28]

The integration is often transparent (meaning, in the weird way that word works, that you can't see the machinery that links one service to another company). But it enables the sharing of powerful functionality across many different sites. Not only does everyone not have to reinvent the wheel. They also don't have to build it. Web services enable the invention and the building to be shared among many different entities.

This is a pattern that will grow dramatically as more companies follow the same path. When you go to a blog, for example, the comments might be handled by a special comment company (necessitated by evil spammers). Or when you answer a poll at a Web site, some other Web site will actually be running the poll. But visible or not, the effect will be quite profound. These technologies will radically reduce the cost of doing business in this increasingly important commercial space.

LEGO-ized innovation is just one component of what Tim O'Reilly first tagged "Web 2.0."[29] It may ultimately be the most important. For it demonstrates both how the Internet is uniquely poised to exploit a general tenet of economics and how the Internet takes advantage of the principle of democratization that is its hallmark. Consider these two in turn.

Economics

In 1937 Nobel laureate Ronald Coase was wondering why there were firms in a free market.[30] If the core of a market was that resources should be allocated by price, why within a firm wasn't it price that determined who got what? Within a firm it was the command of a "boss." Life inside the firm thus looked more like the "economic planning" of communism than the competition of a marketplace. Why? Why weren't firms built like free markets?

The answer was "transaction costs." It cost money to go to the market: time, bargaining costs, costs of capital, etc. Coase reasoned that this cost would help explain the size of a firm. A firm would go to the market to obtain a product when doing so was cheaper than producing the product inside the firm. It would produce the product in house when the costs of the market were too high. Yochai Benkler summarizes the point:

> [P]eople use markets when the gains from doing so, net of transaction costs, exceed the gains from doing the same thing in a managed firm, net of the costs of organizing and managing a firm. Firms emerge when the opposite is true, and transaction costs can best be reduced by bringing an activity into a managed context that requires no individual transactions to allocate this resource or that effort.[31]

It follows from this insight that as transaction costs fall, all things being equal, the amount of stuff done inside a firm will fall as well. The firm will outsource more. It will focus its internal work on the stuff it can do best (meaning more efficiently than the market).

LEGO-ized innovation is simply the architectural instantiation of this economic point. Through the architecture that makes Web 2.0 possible—including what many have called Web services—the transaction costs of outsourcing functionality drop dramatically. Why set up a payment service—exposing yourself and your firm to the risk of fraud, for example—when you can simply contract with PayPal? Why run your own servers when a firm can really promise 24/7 service with its own? Some realms, like national security, might well want to opt out of this sort of outsourcing. But the obvious point is that it will make sense not to outsource less and less.

Democratization

LEGO-ized innovation also teaches us something critical about innovation on the Internet itself. In each of these Web 2.0 examples, the platform allows innovation to be, as MIT professor Eric von Hippel describes, "democratized." Once again, that term does not mean innovation gets implemented as strategy was decided in the early brigades of Soviet soldiers—by gathering around and voting on the next strategic move. Instead, "democratized" here means that access to the resource—the right to innovate—has been made more *democratic,* that is, made dependent upon your membership in some community, and not upon a special status or hierarchy within some company or government.

Amazon and Google democratize innovation when they open their Web services to people outside Amazon and Google. The Internet did the same, just better. The original architecture of the Internet was called "end-to-end," meaning innovation and intelligence in the network were to be at the edge of the network (the machines that connect to the network, not the network itself); the network itself was to be as simple as it could be.[32] As a result, anyone was technically free to innovate for this network. All you needed to do to innovate for the Internet was to conform your design to the Internet's protocols. Once you did that, you were in. There was no committee or design group or Agency of Internet Innovation that needed to approve your idea. Nobody could stop you from building whatever you wanted to build on the Internet. That freedom is a critical reason for the Internet's extraordinary success.

The Character of Commercial Success

You can tell a great deal about the character of a person by asking him to pick the great companies of an era. Does he pick the successful dinosaurs? Or does he pick the hungry upstarts?

My taste is for the hungry upstarts. One great feature of modern society is the institutionalized respect we give to processes designed to destroy the past. The free market is the best example. Democracy is another. In both cases, constant flux is not the objective (we have courts to protect private property; we have constitutions to slow the will of the democracy). But in both cases, the aim is to assure that the past survives only if it can beat out the future.

The commercial economies of the Internet are a fantastic example of exactly this dynamic. The neutral platform of the Internet democratized technical and commercial innovation. Power was thus radically shifted. The dropouts of the late 1990s (mainly from Stanford) beat the dropouts of the middle 1970s (from Harvard): Google and Yahoo! were nothings when Microsoft was said to dominate. This success of the new against the power of the old was made possible by a constitutional commitment in the architecture of the network to democratize innovation.

No government could have planned these successes, and not just because governments are unlikely to have the talent of the geniuses at the likes of a Google or an eBay. Rather, governments couldn't plan these successes because governments, at least as we Americans know them, are inherently corrupted—not by bribery, not by greed, but by the reality of campaign financing, which lets them understand the views of only the last great success, and never the views of the next great success (which, as yet, lacks the funds to influence the government).

Nor did these successes come from the dominant business of the time: Amazon beat (the more established) Barnes & Noble. Netflix beat (the innovator) Blockbuster. And Apple beat Dell. That's not because Barnes & Noble was stupid and Amazon was smart.

Rather, as Clayton M. Christensen put it in his justly acclaimed book, *The Innovator's Dilemma:*

> Despite their endowments in technology, brand names, manufac-
> turing prowess, management experience, distribution muscle, and
> just plain cash, successful companies populated by good managers
> have a genuinely hard time doing what does not fit their model for
> how to make money. Because disruptive technologies rarely make
> sense during the years when investing in them is most important,
> conventional managerial wisdom at established firms constitutes
> an entry and mobility barrier that entrepreneurs and investors can
> bank on. It is powerful and pervasive.[33]

Smart for one time does not translate into smart for the next time. For that, we need new businesses.

The Long Tail, Little Brother, and LEGO-ized innovation explain part of the success of the Internet economy. They explain why commerce in the Internet economy can function better (that is, more efficiently) than commerce in real space.

Yet not all of the value from the Internet comes from this commercial economy. Indeed, a more surprising source has nothing to do with commerce at all. It is to that part we now turn.

Sharing Economies

Sitting next to me on a cross-country flight was a representative of America's youth. He was about seventeen, dressed in a complicated mix of black and silver (the metal, not the color). He had a computer far cooler than mine. And when the chime indicated that "it is now safe to use approved electronic devices," he pulled from the seat pocket in front of us a huge portfolio of DVDs.

All of them—there must have been two hundred at least—were copies. And as he paged through the binder, my envy grew. I wanted to know more about his collection and him. So I did something simply awful, something that I never do: I struck up a conversation with the person sitting next to me on an airplane.

I asked Josh (it turned out) about his collection. Was he a film studies student? Did he work in the industry? He wasn't. And he didn't. He was just a collector. Indeed, a collector of "everything," he told me. This was just part of his collection. He had "gigs" of music as well.

The more we spoke, the more conflicted I became. I admired his knowledge. He knew his culture better than I knew mine. But he was, according to the laws of our country, a thief. Or something like that. In building his collection, he had violated a billion rights. Don't start with me about how those rights are unjustly framed, or too expansive, or outdated. I know all that. I've killed forests explaining all that. All that aside, what this kid was doing was making my work harder. I fight for "free culture." My position is weakened by kids who think all culture should be free.

When the frustration of the conflict became too much, I looked for an easy escape. Josh had a film I had always wanted to see. My book was finished. My e-mail was just annoying. I decided I'd ask to watch one of his DVDs.

"So," I said, "could I rent one of those from you? How about $5?"

I'm not writer enough to describe the look of utter disappointment on his face. Suffice it to say that I had found the single most potent insult to hurl at Josh.

"What the fuck?" he spit back at me. "You think I do this for money? I'm happy to lend you one of these. But I don't take *money* for this."

I had crossed a line. But with that crossing, my respect for Josh grew. I didn't agree with how he had acquired his collection. Yet his rebuke reminded me of a different economy within which culture also lives. There exists not just the commercial economy, which meters access on the simple metric of price, but also a sharing economy, where access to culture is regulated not by price, but by a complex set of social relations. These social relations are not simple. Indeed, these relations are insulted by the simplicity of price. And though I hope not many trade on capital acquired as Josh acquired his, everyone reading this book has a rich life of relations governed in a sharing economy, free of the simplicity of price and markets.

If the point isn't completely obvious, consider some more examples:

- You have friends. That friendship lives within a certain economy. If you only ever ask and never give, the friendship goes away. If you meter each interaction and demand a settlement after each exchange, the friendship also goes away. Certain moves appropriate in some places are inappropriate here. For example: "I need to talk to someone. Can I give you $200 for an hour-long session?"
- You have, or have had, or will have, lovers. That relationship exists within a complex, sharing economy. The statement "Wow, that was great. Here's $500!" isn't gratitude in such

relationships. It might be perversion, though if not matched
by perversion on the other side, it will likely be terminal to
the relationship. Lovers make demands on each other. Those
demands are designed to be complex. Simplify them according to price, and you destroy the relationship. (The other
side to this story follows directly as well: Prostitution is sex
within a commercial economy. Both sides seek the simplicity of cash. Crossing that boundary is the stuff of novels or
career-launching movies [Julia Roberts, *Pretty Woman*].)

- You have neighbors. They (or you) will sometimes need help.
  Once one asked me: "My car battery is dead. Can you give me
  a jump?" After we got his car started, he tried to hand me
  $5. "What the hell, Ted," I said. "This is what neighbors do."
  Then I thought, but didn't say: Anyway, if you were going to
  pay me for this hassle, it's going to be a lot more than $5.

As with any economy, the sharing economy is built upon exchange.
And as with any exchange that survives over time, it must, on balance, benefit those who remain within that economy. When it
doesn't, people leave. Or at least they should (think about the bat-
tered spouse).

But of all the ways in which the exchange within a sharing
economy can be defined—or put differently, of all the possible
terms of the exchange within a sharing economy—the one way in
which it cannot be defined is in terms of money. As Yochai Benkler
puts it, in commercial economies "prices are the primary source of
information about, and incentive for, resource allocation"; in sharing economies "non-price-based social relations play those roles."[34]

Indeed, not only is money not helpful. In many cases, adding
money into the mix is downright destructive.[35] This is not because

people are against money (obviously). It is instead because, as philosopher Michael Walzer has described generally, people live within overlapping spheres of social understanding. What is obviously appropriate in some spheres is obviously inappropriate in others.[36]

Both academic literature and ordinary life are filled with a rich understanding of the differences between commercial and sharing economies. My favorite is Lewis Hyde's *The Gift*, which describes in great historical detail the different but related understandings that cultures have had about giving. Think, for example, about the term "Indian giver," which I always understood to be derogatory. It meant someone who gave something but expected to take it back. But the origin of the term invokes the idea of a sharing economy directly—not that you will take the same thing back, but that you understand you're part of a practice of exchange that is meant, over time, to be fair: "In 1764, when Thomas Hutchinson wrote his history of the colony, the term was already an old saying: 'An Indian gift,' he told his readers, 'is a proverbial expression signifying a present for which an equivalent return is expected.'"[37] So why then do people give such gifts, the man from Mars asks? Why do they risk the gift's misfiring? Why not simply give cash, which is guaranteed to transfer efficiently?

The answer is because the gift is doing something more, or different, from simply transferring an asset to another. Again, as Hyde describes it:

> It is the cardinal difference between gift and commodity exchange that a gift establishes a feeling-bond between two people, while the sale of a commodity leaves no necessary connection. I go into a hardware store, pay the man for a hacksaw blade and walk out. I may never see him again. The disconnectedness is, in fact, a virtue

of the commodity mode. We don't want to be bothered. If the
clerk always wants to chat about the family, I'll shop elsewhere. I
just want a hacksaw blade.[38]

Gifts in particular, and the sharing economy in general, are thus
devices for building connections with people. They establish relationships, and draw upon those relationships. They are the glue of
community, essential to certain types of relationships, even if poison
to others. It is not a gift relationship that defines your employment
contract with a steel mill. Nor should it be. But it is a gift relationship, or sharing economy, that defines your life with your spouse
or partner. And if it isn't, it better become so if that relationship is
to last.

Sometimes organizations trade upon this kind of economy in
order to trade upon the kind of connections a sharing economy
produces. Hyde points to the extraordinarily successful example of
Alcoholics Anonymous:

> AA is an unusual organization in terms of the way money is
> handled. Nothing is bought or sold. Local groups are autono-
> mous and meet their minimal expenses—coffee, literature—
> through members' contributions. The program itself is free. AA
> probably wouldn't be as effective, in fact, if the program was
> delivered through the machinery of the market, not because its
> lessons would have to change, but because the spirit behind them
> would be different (the voluntary aspect of getting sober would
> be obscured, there would be more opportunity for manipulation,
> and—as I shall argue presently—the charging of fees for service
> tends to cut off the motivating force of gratitude, a source of AA's
> energy).[39]

Likewise, communities that were defined as sharing economies radically change when money is brought into the mix. Hyde quotes MIT geneticist Jonathan Kind:

> In the past one of the strengths of American bio-medical science was the free exchange of materials, strains of organisms and infor-mation. But now, if you sanction and institutionalize private gain and patenting of micro-organisms, then you don't send out your strains because you don't want them in the public sector. That's already happening now. People are no longer sharing their strains of bacteria and their results as freely as they did in the past.[40]

In all these cases, price is poisonous. Money changes a relation-ship—it redefines it. Indeed, it would most likely insult the host. "Money-oriented motivations are different from socially oriented motivations."[41] And crossing the line will either show a profound misunderstanding of the context, or suggest you did understand the context, but simply wanted to change it.

These lines of understanding, of course, are not drawn by God. They are culturally and historically contingent. In Victorian England, for example, "the presence of money in sport or enter-tainment" reduced the value of that sport or entertainment, at least for "members of the middle and upper classes."[42] Obviously, Americans feel differently today. In nineteenth-century America, the idea that you would tell your personal problems to a paid pro-fessional would seem outrageous. Today, it is called therapy—and the phrase "hey, save that one for the couch" signals an increas-ing appreciation that some personal matters are not to be within a sharing economy. Some personal matters should simply be professionalized.

Thus, no distinction between "sharing" and "commercial" economies can be assumed to survive forever, or even for long. My only claim is that when such a distinction exists, then "adding money for an activity previously undertaken without price compensation reduces, rather than increases, the level of activity."[43] *Often,* not always. Conservatives in America insist upon keeping prostitution illegal because they fear that adding money to sexual exchange will increase the "activity previously undertaken without price compensation"—i.e., sexual activity outside a monogamous relationship. In that case, the fear is money increases the activity, not decreases it.

Commercial and sharing economies coexist. Indeed, they complement each other. Psychologists don't begrudge friendship, even though the stronger the economies of friendship in a society, the weaker the demand for shrinks. The band Wilco doesn't begrudge a church choir, even if the choir gives its work away for free, while Wilco charges plenty for one of their (too infrequent) concerts. We all understand that similar things can be offered within different economies. We celebrate this diversity. Only a fanatic would advocate wiping away one economy simply because of its effect on the other.

Yet sometimes we're all fanatics. Puritan society has waged war against economies for sex that compete with sex within a monogamous relationship—believing both fornication (a competing sharing economy) and prostitution (a competing commercial economy) put too much pressure on an idealized sharing economy. Likewise, the content industry today wages war against economies for exchanging copyrighted content—peer-to-peer sharing economies, where people don't necessarily know one another, as well as friendship sharing economies,[44] where they do. In both cases, the judgment that the one economy is poison to the other may well be right. But whether right or not in a particular case, the key is that these fanatical cases are the exception. In the vast majority of cases, we permit this intereconomy competition to flourish. In many cases, we encourage it. No one is called a communist because he plays in a Thursday-evening softball league (competing with professional baseball) or helps clean up at a local church (competing with the janitor of the church). To the contrary: we idealize one who can trade within a range of societies, with a significant part of his or her life outside the society of commerce.

Now consider a distinction among the possible motivations that might explain participation within a sharing economy. Sometimes these motivations are "me-regarding"—the individual participates in the sharing economy because it benefits him. Sometimes these motivations are "thee-regarding"—the individual participates in the sharing economy because it benefits others. So if I join a local softball league, I may be driven largely by me-regarding motivations. If I volunteer at a local soup kitchen, I'm probably driven mostly by thee-regarding motivations.

Obviously, me and thee motivations are not unrelated. One can always view motivations that are thee-regarding as being ultimately me-regarding—I choose to help my neighbors because I want to be, or I want to be seen as, the sort of person who helps my neighbors. That's a perfectly sensible way to understand the vast majority of thee-regarding motivations. My aim is not to insist that sharing economies are economies of selflessness.

Yet even if thee-regarding motivations are ultimately me-regarding, they are still, in one sense, more complicated to explain than the simple me-regarding motivations we all understand intuitively. We're tolerant of weird me-regarding motivations (we call some "fetishes," others simply "taste"). But weirdness about thee-regarding motivations makes us wonder whether the person even understands what he's saying. For example, I understand the statement "I'm working to spread the goodness of the National Rifle Association." I understand it even though I wouldn't do the same. But the statement "I'm working to spread the goodness of Exxon" is not just unusual. For anyone not actually employed by Exxon, we'd wonder whether the person really understood what he was saying. Thee-regarding motivations plug into existing understandings of communities or causes. Me-regarding motivations (for us, in modern tolerant societies) aren't so constrained.[45]

Using this distinction, then, I will call "thin sharing economies" those economies where the motivation is primarily me-regarding; "thick sharing economies" are economies where the motivations are at least ambiguous between me and thee motivation. Thus, in thin sharing economies, people do not base an exchange on price or money. But they're making this exchange simply because it makes them better off, or because it is an unavoidable by-product of something they otherwise want to do for purely me-regarding reasons. One person doesn't necessarily mind that his actions might be helping someone else. But there's no independent desire to help someone else. The motivation is about me.

Three examples will illustrate what I mean.

- Think about a stock market. In most major stock markets, people share information—ordinarily information about how much is bought at what price, but even if that were hidden, the market would share the information about how prices were changing. You could describe this sharing as constituting a sharing economy. But plainly, it's a very weird sort of person who would buy and sell stocks simply to help the market collect information about prices. People buy and sell stocks to make money. A by-product of that behavior is the information that gets shared with others. If this is a sharing economy, it is a thin sharing economy.

- Think of the "Voice Over IP" service called Skype. With Skype, you can make free Internet calls, and very cheap Internet-to-regular-phone calls (and vice versa). But Skype is designed to use, or "share," the resources of the computers connected to this VOIP network. When you're on the Skype phone, Skype is using your computer to make its network work better.[46] This is like AT&T drawing electricity from your house when you use the telephone, as a way to keep its electricity costs down. I don't mean to criticize Skype for this: it certainly helps make the service better. But when someone participates in this "sharing economy" of computer resources, what is the most salient motivation? Is it to advance the cause of Skype? Or is it simply a by-product of people's desire for cheap calls? I suggest the latter, making this too a thin sharing economy.

- Think finally of AOL's IM network. The value of that network increases for everyone. This is a consequence of network effects: the more who join, the more valuable the resource is for everyone. There are many contexts in which this network effect is true. Think, for example, about the English language. Every time someone in China struggles to learn English or a school in India continues to push English as a primary language, all of us English speakers benefit. But in neither of these cases—with AOL or English—are people joining the movement because it is a movement. People join because it gives them something they want.

In each case, there is a resource that is shared among everyone within the community—information about the market, computer resources to make VOIP work better, the network effect from a popular network. That resource is shared independent of price. But in none of these cases is it realistic to imagine people joining or participating in these networks for thee-regarding reasons. These are me-regarding communities. They are thin sharing economies.

By contrast, in a thick sharing economy, motivations are more complex. A father might spend Sunday mornings teaching a Bible class at his church. Part of that motivation is about him. But certainly, part is also about improving the community of his church—a thee motivation. What the proportion is we need not specify. The only important point is that there are both, and that the more we think that there is a thee motivation, the thicker the community is.

This distinction between thick and thin will be important when considering differences among sharing economies. It will also be important in understanding the likelihood that any particular economy will survive over time. For despite the intuitions that names give to the contrary, a thin sharing economy is often easier to support than a thick sharing economy. This is because inspiring or sustaining thee motivations is not costless. Or at least, all things being equal, a me motivation (for us, now) comes more easily to most. Thus, distinguishing cases where a thee motivation is necessary from cases where it isn't will be helpful in predicting whether a certain sharing economy will survive.

Internet Sharing Economies

The Internet has exploded the range and thickness of sharing economies too. As with commercial economies, the plasticity of the Internet's design, and the scale of its reach, offer a vast range of new opportunities for sharing economies everywhere.

As with commercial economies, these sharing economies flourish in part because of their design. Here too, for example, the best follow a Bricklin-like principle: People contribute to the common good as a by-product of doing what they would otherwise want to do. But some communities demand something more from their members: some will claim, for example, that members owe one another something. Depending upon the community, that demand will often stick. If you told me I had a duty to Amazon, I'd think it a joke. I love Amazon as much as the next guy. But it gets no loyalty beyond the good that it offers in return. But there are plenty of entities within the Internet sharing economy for whom it isn't a joke to say I owe the community something. The best such communities may not depend upon this kind of owing. They may simply make doing good fun. But in some communities, all the participants understand they must "do their part." And failing to do his part opens the deviant to criticism. For these thick sharing economies, the motivations to participate are more complex.

The most prominent Internet sharing economy today, and a paradigm of the type, is one that didn't even exist before 9/11: Wikipedia. But Wikipedia is not the first Internet sharing economy. So after we cover the familiar and dominant, we'll go backward a bit, to better appreciate the continuity between the "barn raising," as one of the Net's early legal theorists, Mike Godwin, put it, of Wikipedia, and the many barn raisings that happened before Wikipedia was born.

The Paradigm Case: Wikipedia

In 2000, Jimmy "Jimbo" Wales was fishing around for something better to do. He had been a futures and options trader in Chicago during most of the 1990s and had made, he told *Wired* magazine, enough money "to support himself and his wife for the rest of their lives."[47] Now he wanted to do something really interesting.

At first he thought about writing an encyclopedia, or at least getting an online encyclopedia written. Using some of the profits from an adult-content site that he had helped start (Bomis), Wales launched Nupedia. The idea—obviously the only sane idea for writing an encyclopedia at the time—was to build a peer-reviewed work. He hired a philosophy Ph.D., Larry Sanger, as editor in chief. And they both watched this pot as the project never boiled.

Frustrated over its slow growth, Nupedia launched a "wiki" to encourage the development of Nupedia articles. A wiki is a platform that lets anyone write or edit in a common space. Wiki software has been around for more than a decade. It was originally intended to enable a team to work on a project collaboratively. Wales and Sanger intended the wiki to be a sandbox for collaborative drafting of articles for Nupedia. Quickly, however, the sandbox became much more than a draft. The growth of articles in this (now dubbed) "Wikipedia" dwarfed anything on Nupedia. The sandbox then took center stage.

Wikipedia, however, is more than software. It is also a set of norms that were built into the practice of using that software. The objective was an encyclopedia. That meant articles were to be written from a "neutral point of view" (NPOV). And the project was to be run by a volunteer community (though Sanger was originally a paid editor so long as Bomis's funding continued). To assure that the volunteers felt they were part of a community, the rules had to be rules anyone could live by. Thus was born the "ignore all rules" rule, which Jimmy Wales explained to me as follows:

> "Ignore all rules"…is not an invitation to chaos. It is really more an idea of saying, "Look, whatever rules we have in Wikipedia, they ought to be, more or less, discernible by any normal, socially adept adult who thinks about what would be the ethical thing to do in this situation. That should be what the rule is." It should be pretty intuitive. And if there's something that's counterintuitive, it shouldn't really be a rule. It might be a guideline or it might be something that we go around and try to encourage people to do. But you can't get in trouble for not doing it.[48]

Finally, there was a norm about ownership: nobody owned Wikipedia exclusively. The content of Wikipedia got created under a copyright license that guaranteed it was always free for anyone to copy, and that any modifications had to be free as well. This "copyleft" license—the brainchild of Richard Stallman—set the final founding norm for this extraordinary experiment in collaboration.

If you're one of the seven people in the world who have not yet used Wikipedia, you might well wonder whether this experiment in collaboration can work. The answer is that it does, and surpris-

ingly well—surprising even for Wikipedia's founder, Jimmy Wales.
As he explained to me:

> As people get experienced using Wikipedia and they're reading
> it a lot, they begin to have this intuition that Wikipedia is pretty
> darn good about being neutral on very controversial subjects. And
> that's a little bit surprising; I know certainly if you had asked me
> before Wikipedia what a big problem would be, I would have
> said, "Wow, I'm hoping that it's not going to be incredibly biased
> on controversial subjects. I'm hoping that that won't happen." It
> turns out that doesn't happen, that community is quite good … in
> part because of the social norm that we've had from the beginning
> about neutrality and about communication.

Not all of the work within Wikipedia is writing original articles.
Indeed, the vast majority of work is editing content—correcting
spelling or formatting errors, rewriting submissions to conform
to the NPOV norm, or simply "softening [a claim] to be more
broadly acceptable." According to one estimate, only 10 percent
of all edits add substantive content.[49] The rest is cleaning up those
additions. And even here, more of the work is done by a relatively
small number of users. According to Jimmy Wales, 50 percent of
all edits are done by 0.7 percent of users—meaning just about 524
users within his sample. The most active 2 percent (1,400) of users
have done 73.4 percent of all edits. Counting content, Aaron Swartz
found that "the vast majority of major contributors are unregis-
tered and that most have only made a handful of contributions to
Wikipedia."[50]

This division of work is not directed. There's no "chore" norm
at Wikipedia. As Wales describes,

If somebody says, "Well, I know about birds and I'm going to come in and monitor a few hundred bird articles and I'm going to occasionally update them when I feel like it but I'm in and out and I'm not really a core community member. And I, frankly, don't really have time or feel like dealing with the conflict and I'm not going to run a spell-checking bot and I'm just going to do the parts that I find fun," that's considered perfectly acceptable.

These are volunteers doing as they like. It just turns out that when you invite the world to participate, there are enough volunteers in a range of categories of work to make the whole thing function quite well.

The first question many ask about these thousands of volunteers is, why do they do it? (And again, this is a world of volunteers. Until February 2005, there was just one part-time employee).[51] "Why do people play softball?" is a standard Wales response.[52] The answer of course is simply because they like it more than all the other things they might be doing at the time. But *why* do they like it? In part because there is also a ready, and attractive, thee-regarding motivation surrounding the project. As Wales told Tapscott and Williams, "We are gathering together to build this resource that will be made available to all the people of the world for free. That's a goal that people can get behind."[53]

That goal makes Wikipedians (as they call themselves) a community—not in some abstract sense of a bunch of people with a common interest, but instead in the very significant sense of people who have worked together on a common problem. As Wales describes,

Community sometimes is almost meaningless; it just means there's people out there doing stuff. But in Wikipedia, what community means is that they're people who have met each other; they know each other; they've had arguments; they've made up; they've had different kinds of controversies; they've banded together to take care of some problems; they like each other; they don't like each other; sometimes people are dating and then they break up and then there's some rumors and scandals, and all of the stuff that makes a rich human community is what goes inside Wikipedia. It's a complete soap opera actually inside our community.

These people are likely to pick up any litter they see in their streets.

Surprisingly, Wikipedia is even good at things you wouldn't associate with a traditional encyclopedia—reporting and analyzing news events such as the Virginia Tech massacre and Hurricane Katrina. Wales explains:

One of the things that we are doing better, I think, is when we have a mass public event or story with breaking news, one of the things that we've seen is that, in the short run, especially, Wikipedia does a very interesting thing that I have come to appreciate more and more over time, which is a census of the news that's coming out. So, the way I present this is when you have a big event like this, you'll have ten, twenty, or thirty, or fifty reporters all there, on the scene gathering information. But they're each seeing only the piece that they can see and even if they're all absolutely excellent journalists who are doing their very best to get the whole story, they're each coming from a particular perspective and they're each interviewing particular people with particular views. And then that stuff goes out onto the Web where people can read all of it.

The *New York Times* made the same point after the Virginia Tech massacre. As a review article noted, "From the contributions of 2,074 editors, at last count, the site created a polished, detailed article on the massacre, with more than 140 separate footnotes, as well as sidebars that profiled the shooter, Seung-Hui Cho, and gave a timeline of the attacks."[54] That article was viewed by more than 750,000 within the first two days. Even the local newspaper, the *Roanoke Times,* commented that Wikipedia "has emerged as the clearinghouse for detailed information on the event."[55]

I've called Wikipedia part of the "sharing economy" even though technically the license governing Wikipedia permits anyone to copy Wikipedia for whatever purpose he or she wants, including the purpose of selling copies. There's nothing wrong, according to the license at least, with running an ad-supported site with a copy of Wikipedia. There's no problem in printing a physical copy of the hundred most popular articles and selling those copies for money. The only licensing restriction is that if you make changes to Wikipedia, you have to license the new version under the same license as the old. No one is permitted to improve and then lock up the improvements. They too must remain free.

But Wikipedia is still part of the sharing economy because one's access to, or right to edit for, Wikipedia is not metered by money. More interestingly, the site itself—the one owned by the Wikimedia Foundation—doesn't run ads to support its costs. That decision is extremely significant. As one of the top ten Web sites in the world, the decision not to run ads means Wikipedia leaves about $100 million on the table every year. Why? What drives this site to ignore so much potential wealth?

One reason important to Wales relates directly to the importance of the NPOV. As he explained to me, "We do care that the general public looks to Wikipedia in all of its glories and all of its flaws, which are numerous of course. But the one thing they don't say is, 'Well, I don't trust Wikipedia because it's all basically advertising fluff.'" Forgoing ads is a way to buy credibility, just as a judge forgoing bribes is a way to buy credibility. In both cases, we might imagine the entity taking money would not be affected by that money. But there's no easy way to verify that it's not been affected. So to achieve the value sought—neutrality, or fairness—money must be removed from the equation.

Wikipedia is my paradigm sharing economy. Its contributors are motivated not by money, but by the fun or joy in what they do. Some find that joy because the result is something valuable to society. Others find that joy because there's nothing better on television. Whatever the reason, there's sufficient motivation spread throughout the world to build an encyclopedia for free that each day draws more attention than all the other encyclopedias in history combined. Wikipedia is to culture as the GNU/Linux operating system is to software: something no one would have predicted could have been done, yet which an inspired leader and devoted followers built for free, and to remain free.

Beyond Wikipedia

The Internet learned to share, however, long before Wikipedia. Indeed, as commerce was banned from the Internet until 1991, one might well say that the Internet was born a sharing economy; commerce was added only later. There are many examples. Consider just a few:

• *The code that built the Net came from a sharing economy.* The software that built the original Internet was the product of free collaboration. Open-source, or free, software was distributed broadly to enable the servers and Internet protocols to function. The most famous of these projects was the GNU Project, which in 1983 was launched by Richard Stallman to build a free operating system, modeled upon the then dominant UNIX. For the first six years or so, Stallman and his loyal followers worked away at building the infrastructure that would make an operating system run. By the beginning of the 1990s, the essential part missing was the kernel of the operating system, without which the operating system as a whole could not run.

A Finnish undergraduate decided to try to build that kernel. After tinkering a bit with a version, he released it to the Net for others to add to. This undergraduate was named Linus Torvalds. He named the kernel Linux. Soon, volunteers from around the world had helped improve the kernel enough that, when added to the other components of the GNU system, it built a robust and powerful operating system called either Linux or, better, GNU/Linux. We'll see more about this operating system in the next chapter. The point to remark upon here is it was built by thousands volunteering to write code that would eventually guarantee that people could build upon and share an operating system.

Less famous than GNU/Linux, but just as important to the history of the Net, are the many instances of free software built to supply the basic plumbing of the Internet. As Robert Young and Wendy Goldman Rohm put it in their book, *Under the Radar* (1999):

In 1981, Eric Allman created Sendmail, an open source program that is responsible for routing 80 percent of the email that travels

over the Internet. It is currently still maintained by thousands of
online programmers via sendmail.org. In addition, Allman started
Sendmail Inc. as a business in November 1998. For a profit, he
sells easy-to-use versions of the open source software, along with
support and service, to corporations. Another important force in
the open source world is Perl. It was created by 43-year-old Larry
Wall, a former linguist who created Perl while at Burroughs Corp.
on a government-funded project. The software is free, although
Wall has sold 500,000 copies of his Perl manuals. Another open
source program, BIND, was originally developed at the University of California at Berkeley as freeware. It allows domain names
like Linux.com to be entered as textual name addresses instead of
machine numbers (called IP addresses, for example, 43.72.66.209),
making it much easier for ordinary people to surf the Internet.
Apache, the group founded by 25-year-old Brian Behlendorf, got
its start when Behlendorf was hired to build *Wired* magazine's
Web site. In order to improve the Web server software, he programmed his own enhancements and circulated the results, with
source code, on the Internet. Other contributors added their code,
and Apache was created. The name came from the fact that the
software was "a patchy" collection of code from numerous contributors. Currently, Apache is used by more than half of the Web
sites on the Internet. It was chosen by IBM, over Netscape's and
Microsoft's closed-source Web server software, to be the foundation of IBM's Web commerce software.[56]

Apache continues to be the dominant Web server on the Internet: for most of the first half of the decade, its market share was
over 60 percent; today, despite fierce competition from proprietary

server companies such as Microsoft and Apple, the market share remains in the mid-50-percent range.[57] All of these products were initially built by people who lived within an economy of exchange. But their interactions within that economy were not metered by money. Some were paid by others so that they could afford to write software that would be free. But the terms of exchange for adding and changing this code were forbidden to be commercial. The core free-software license permits developers to sell their code. But they can never sell the right to modify or change the code they build onto free software. That economy is always to be a sharing economy.

Why does this kind of software development work? Or better, why does it often work so much better than proprietary software?

One reason is structural: when you write software that others are to work on, you must be more disciplined in your coding. Comments must be frequent. Code must be made more modular. That structure helps evaluate bugs. It also invites more to review the work of the coder: "with enough eyeballs all bugs are shallow."[58]

But there's a third reason that is frequently ignored. Free and open-source software takes advantage of the returns from diversity in a way that proprietary software hasn't. As economist Scott Page has demonstrated in a foundational study about the efficiency of diversity, the success of an enterprise in solving a difficult problem depends not just upon the *ability* of the people solving the problem.[59] Using mathematical economics, Page shows that the success also depends upon the *diversity* of the people solving the problem. What's needed is not just, or even necessarily, racial diversity, but a diversity in experience and worldviews, so as to help a project fill in the blind spots inherent in any particular view.

That point in the abstract might not sound surprising: sure, diversity helps, just like ability helps. But the really surprising part of Page's analysis is the relationship between the contribution from ability and the contribution from diversity: equal. Increasing diversity, in this sense, is just as valuable as increasing ability.

Thus, between two projects, one in which the workers are extremely smart but very narrow, and another in which the workers are not quite as smart but much more diverse, the second project could easily outperform the first. So even if you believe that proprietary firms can hire the very best programmers, an open-source project (with a wider diversity of coders) could easily outperform the proprietary project.

This dynamic, I suggest, explains a great deal of the success of the software sharing economy. It likewise could explain the success of Internet sharing economies as well.

• *Project Gutenberg is a sharing economy.* Founded in 1971 (yes, *1971*), Project Gutenberg is the oldest digital library. Its founder, Michael Hart, launched the project to digitize and distribute cultural works. The first Project Gutenberg text was the Declaration of Independence. Today, there are more than twenty-two thousand books in the collection, with an average of fifty books added each week.[60] The vast majority of the books in the collection are public-domain works, primarily works of literature. Most are in English, and most are available in plain text only.[61] Hart describes his mission quite simply: "to encourage the creation and distribution of e-books." The economy of Project Gutenberg is a sharing economy. Volunteers add works to the collection; people download works freely from the collection. Price, or money, doesn't police access.

Voluntary contributions are all the supporters can rely upon to keep the work alive.

• *Distributed Proofreaders is a sharing economy.* Inspired by Michael Hart's Project Gutenberg, and launched in 2000 by Charles Franks, the Distributed Proofreaders project was conceived to help proofread for free the books that Hart made available for free. To compensate for the errors of optical character recognition (OCR) technology, the Distributed Proofreaders project takes individual pages from scanned books and presents them to individuals, along with the original text. Volunteers then correct the text through a kind of distributed-computing project. (See the next item for more on distributed computing.) Distributed Proofreaders has contributed to more than ten thousand books on Project Gutenberg. In 2004, there were between three hundred and four hundred proofreaders participating each day; the project finished between four thousand and seven thousand pages per day—averaging four pages every minute.[62] All of this work is voluntary.

• *Distributed-computing projects are sharing economies.* Distributed computing refers to efforts to enlist the unused cycles of personal computers connected to the Net for some worthy cause (worthy in the eyes of the volunteer, at least). The most famous was the SETI@home project, launched in 1999 and designed to share computing power for the purpose of detecting extraterrestrial life (or at least the sort that uses radios). More than 5 million volunteers eventually shared their computers with this project.[63] But there are many more distributed-computing projects beyond the SETI project. A favorite of mine is Einstein@Home. As described by Wikipedia,

Einstein@Home is designed to search data collected by the
Laser Interferometer Gravitational-Wave Observatory (LIGO)
and GEO 600 for gravitational waves. The project was officially
launched on 19 February 2005 as part of American Physical Society's contribution to the World Year of Physics 2005. It uses the
power of volunteer-driven distributed computing in solving the
computationally intensive problem of analyzing a large volume of
data.... As of June 3, 2006, over 120,000 volunteers in 186 countries have participated in the project.[64]

The contributions to these distributed-computing projects are
voluntary. Price does not meter access either to the projects or to
their results.

  • *The Internet Archive is a sharing economy.* Launched in 1996 by
serial technology entrepreneur (and one of the successful ones) Brewster Kahle, the Internet Archive seeks to offer "permanent access for
researchers, historians, and scholars to historical collections that exist
in digital format."[65] But to do this, Kahle depends upon more than
the extraordinarily generous financial support that he provides to the
project. He depends as well upon a massive volunteer effort to identify and upload content that should be in the archive. The archive
employs "probably less than one-tenth of one person," he told me.
And "there have probably been over a thousand people that have
uploaded" creative work to be preserved.[66] All of the content is shared
on the archive. Nothing is metered according to price.

  • *The Mars Mapping Project was a sharing economy.* Scientists at
NASA are eager to map the surface of Mars. Mapping means iden-

tifying and marking on their maps the locations of craters, the age of craters, and other significant geological formations. For years, NASA and others had done this by hiring professionals. For eleven months beginning in November 2000, NASA experimented with asking amateurs to do what professionals had done.

The theory of the experiment was that "there are many scientific tasks that require human perception and common sense, but may not require a lot of scientific training." So NASA set up a site where volunteer "clickworkers" could spend "a few minutes here and there" and some would "work longer" doing "routine science analysis that would normally be done by" a professional.[67] For example, the site included "an interactive interface in which the contributor…clicks on four points on a crater rim and watches a circle draw itself around the rim.…Pressing a button submits the set of latitude, longitude, and diameter numbers to [the] database."[68]

The results were astonishing. Once word of the project got out, there were "over 800 contributors who made over 30,000 crater-marking entries in four days."[69] Even after error correction, this was "faster than a single graduate student could have marked them, and also far faster than the original data was returned by the spacecraft." Thirty-seven percent of the results were provided by onetime contributors. And when the results were redundancy compared, the accuracy was extremely high. As the study of the results concluded, "even if volunteers have higher error rates…, a cheap and timely analysis could still be useful. In some applications, noisy data can still yield a valid statistical result."[70] As Yochai Benkler describes: "What the NASA scientists running this experiment had tapped into was a vast pool of five-minute increments of human judgment, applied with motivation to participate in a task unrelated to 'making a living.' "[71]

• *Astronomy increasingly depends upon a sharing economy.* Historically, astronomy always relied on amateurs. But as digital technologies have made it possible to gather huge amounts of data, there is a strong push within the field to encourage sharing of these data among astronomers. As the editors of *Nature* observed,

> web technologies...are pushing the character of the web from that of a large library towards providing a user-driven collaborative workspace...A decade ago, for example, astronomy was still largely about groups keeping observational data proprietary and publishing individual results. Now it is organized around large data sets, with data being shared, coded and made accessible to the whole community. Organized sharing of data within and among smaller and more diverse research communities is more challenging, owing to the plethora of data types and formats. A key technological shift that could change this is a move away from centralized databases to what are known as "web services."[72]

The limits on this sharing are therefore not technical. They are "cultural." "Scientific competitiveness will always be with us. But developing meaningful credit for those who share their data is essential to encourage the diversity of means by which researchers can now contribute to the global academy."[73]

There's some good evidence this norm is developing. The Digital Sky Project, for example, funded through the National Science Foundation, "provides simultaneous access to the catalogs and image data, together with sufficient computing capability, to allow detailed correlated studies across the entire data set."[74] Likewise with the U.S. National Virtual Observatory, another NSF-funded project, meant to develop "a set of online tools to link all the world's astronomy data together, giving people all over the world easy access to data from many different instruments, at all wavelengths of the electromagnetic spectrum from radio to gamma rays."[75] The emphasis in all these cases is to provide a sharing economy in data, enabling researchers to draw upon the data to analyze and draw conclusions that advance the field of astronomy.

• *The Open Directory Project is a sharing economy.* As a complement to the search algorithms of major search engines, the Open Directory Project "is the largest, most comprehensive human-edited directory of the Web. It is constructed and maintained by a vast, global community of volunteer editors." Volunteers are asked to sign up to a particular area of knowledge. They are given tools to help them edit and modify links within the directory. The directory asks people to give "a few minutes" of their time to "help make the Web a better place."[76] No money polices access to the results of this project, or the right to participate in it.

• *Open Source Food is a sharing economy.* As described by its founder, "Open Source Food came to fruition because me and my father wanted to create a place for people like us. We're not professional cooks, we just love food. We want to share, learn and improve ourselves with the help of like-minded food lovers. Open Source Food is a platform for that." Users contribute recipes to the database of recipes. And while recipes as such can't be copyrighted in the United States, the site uses Creative Commons licenses to make sure descriptive text and images are available freely as well.[77] No money meters access to the site. Contributions are all voluntary.

The list could go on practically indefinitely. The Internet is filled with successful sharing economies, in which people contribute for reasons other than money. As Benkler argues, they contribute not because "we live in a unique moment of humanistic sharing." Rather, the reason for all this sharing is that "the technological state of a society...affects the opportunities for...social, market...and state production modalities."[78] We're living in a time when technology is favoring the social. More vibrant sharing economies are the result.

What Sharing Economies Share

In all of these cases, the people participating in creating something of value share that value independent of money. That's not to say they're not in it for themselves. Nor is it to say that they're not being paid. (A programmer working for IBM may well be paid to add code to a free-software project. But the freedoms that get shared with that free software are not tied to money.) And it's certainly not to say they're in it solely to benefit someone else. All the category of "sharing economy" requires is that the terms upon which people participate in the economy are terms not centered on cash. In each, the work that others might share is never shared for the money.

So why do people do it? What's in it for them? What is their motivation?

This question has been studied extensively in the context of free and open-source software. Its answer begins by recognizing how small the "motivation" is that requires any special kind of explanation. For as a corollary to Dan Bricklin's Cornucopia of the Commons,[79] we need to remember that a large part of the motivation for contributing to these sharing economies comes from people just doing for themselves what they want to do anyway.

With free and open-source software, for example, often the work is self-motivated: a programmer faces a problem and has to fix it ("scratch an itch," as Eric Raymond put it). Eric von Hippel estimates in one study that "Fifty-eight percent of respondents said that an important motivation for writing their code was that they had a work need (33 percent), or a nonwork need (30 percent) or both (5 percent) for the code itself."[80] For these people, the question is not, why does someone write the software? but the much less demanding puzzle, why does someone contribute the solution freely to others? This, as Rishab Ghosh has written, is obviously a simpler problem to explain. You don't lose anything by giving away an intangible good that you've already created; and especially when you've been paid to create it, that's sufficient reason to contribute it to others.[81]

Beyond contributions that are explained by the fact that the contributor had to solve the problem himself anyway, theorists have identified a number of other reasons to explain these contributions to sharing economies. Again with software, one voluntary study demonstrates that a significant portion of contributors are motivated by pure intellectual stimulation (45 percent) or to improve their own programming skills (41 percent listed this as one of their top three reasons.)[82]

Another reason points to a variant on the argument about diversity I identified in Scott Page's work above. As Steven Weber puts it in *The Success of Open Source:*

> Under conditions of antirivalness, as the size of the Internet-connected group increases, and there is a heterogeneous distribution of motivations with people who have a high level of interest and some resources to invest, then the large group is *more* likely, all things being equal, to provide the good than is a small group.[83]

That means developers of open-source and free-software projects have a strong interest in many people sharing the projects, since the more who share them, the more likely someone will be motivated to improve them.

Peter Kollock identifies another potential motivator as the "expectation…of reciprocity. Both specific and generalized reciprocity can reward providing something of value to another. When information providers do not know each other, as is often the case for participants in open source software projects, the kind of reciprocity that is relevant is called 'generalized' exchange."[84]

So we see that there is an abundance, not a lack, of motivations. As Weber writes,

> The success of open source demonstrates the importance of a fundamentally different solution, built on top of an unconventional understanding of property rights configured around distribution. Open source uses that concept to tap into a broad range of human motivations and emotions, beyond the straightforward calculation

of salary for labor. And it relies on a set of organizational struc-
tures to coordinate behavior around the problem of managing
distributed innovation, which is different from division of labor.
None of these characteristics is entirely new, unique to the open
source, or confined to the Internet. But together, they are generic
ingredients of a way of making things that has potentially broad
consequences for economics and politics.[85]

In my view, the easiest answer to the motivation question comes
from framing it more broadly: Why do people do these things for
free rather than, say, watching television?

In some cases, the response is simply that the sharing activity is
more compelling. This is a purely me-regarding motivation. I want
to play a game (MUDs and MOOs), or write an article (Wikipedia), or whatever, because I like to.

In some cases, the response is more thee-regarding: Some part
of the motivation to write for Wikipedia is to help Wikipedia ful-
fill its mission: "Wikipedia is a project to build free encyclopedias in all languages of the world. Virtually anyone with Internet
access is free to contribute, by contributing neutral, cited information." People contribute because they want to feel that they're
helping others. Some people help the Internet Archive or Project
Gutenberg because they want to be part of their mission: to offer
"permanent access for researchers, historians, and scholars to historical collections that exist in digital format" (Internet Archive)
or to "encourage the creation and distribution of eBooks" (Project
Gutenberg).

But again, even the thee-regarding motivations need not be
descriptions of self-sacrifice. I suspect that no one contributes to
Wikipedia despite hating what he does, solely because he believes

he ought to help create free knowledge. We can all understand people in the commercial economy who hate what they do but do it anyway ("he's just doing it for the money"). That dynamic is very difficult to imagine in the sharing economy. In the sharing economy, people are in it for the thing they're doing, either because they like the doing, or because they like doing such things. Either way, these are happy places. People are there because they want to be.

Or more completely, because "they want to be" there given the options the technology offers. As Benkler has put it most clearly, technology doesn't determine any result.[86] But different technologies invite different behaviors. The changes in technology I've described here "have increased the role of [sharing] production."[87] If they continue to grow, they could well become part of the "core, rather than the periphery of the most advanced economies."[88] They have already done much more than anyone would have predicted even ten years ago.

SEVEN

HYBRID ECONOMIES

Commercial economies build value with money at their core. Sharing economies build value, ignoring money. Both are critical to life both online and offline. Both will flourish more as Internet technology develops.

But between these two economies, there is an increasingly important third economy: one that builds upon both the sharing and commercial economies, one that adds value to each. This third type—the hybrid—will dominate the architecture for commerce on the Web. It will also radically change the way sharing economies function.

The hybrid is either a commercial entity that aims to leverage value from a sharing economy, or it is a sharing economy that builds a commercial entity to better support its sharing aims. Either way, the hybrid links two simpler, or purer, economies, and produces something from the link.

That link is sustained, however, only if the distinction between the two economies is preserved. If those within the sharing economy begin to think of themselves as tools of a commercial economy, they will be less willing to play. If those within a commercial economy begin to think of it as a sharing economy, that may reduce their focus on economic reward. Maintaining a conceptual separation is a key to sustaining the value of the hybrid. But how that separation is maintained cannot be answered in the abstract.

The Internet is the age of the hybrid.[1] Every interesting Internet business is now, or is becoming, a hybrid. The reasons are not hard to see. As Yochai Benkler describes,

> A billion people in advanced economies may have between two billion and six billion spare hours among them, every day. In order to harness these billions of hours, it would take the whole workforce of almost 340,000 workers employed by the entire motion picture and recording industries in the United States put together, assuming each worker worked forty-hour weeks without taking a single vacation, for between three and eight and a half years![2]

If sharing economies promise value, it is the commercial economy that is tuned to exploit that. But as those in the commercial economy are coming to see, you can't leverage value from a sharing economy with a hostile buyout or a simple acquisition of assets. You have to keep those participating in the sharing economy happy, and for the reasons they were happy before. For here too money can't buy you love, even if love could produce lots of money.

Yet there are differences among these hybrids. In this chapter, I hope first to hide these differences enough to show a common pattern. I then focus on the differences in order to learn a bit more about how and why some hybrids succeed while others fail. As with Wikipedia among sharing economies, there is a paradigm here too. This is the theme upon which everything else is a variation. I start with this theme and then turn to the growing variations.

The Paradigm Case: Free Software

In the early 1990s, Robert Young was in the computer-leasing business. As a way to bring in customers, he wrote a newsletter called *New York Unix*. The newsletter covered whatever subjects his (potential) customers might be interested in. He was therefore keen to understand precisely what his customers would read. "I would ask all the members of the user groups, 'What do you want to read about that isn't already being covered in the major computer publications?' The only thing they could think about at the time was free software."

So Young decided to learn something about free software. He took a train to Boston to sit down with Richard Stallman to "ask him where this stuff was coming from." Young was astounded by what he found. "[Stallman] was using lines [like] 'from engineers according to their skill to engineers according to their need.'"

"I'm a capitalist," Young recalls thinking, "and the Berlin wall had just fallen. I thought, I'm not sure this model is going to keep going." Young decided to forget about free software. "Given there was no economic support [for this] free software stuff," Young believed it all "was a blip." He reasoned, "It was only going to get worse over time as the communist system only ever got worse over time."

Yet Young quickly saw that like many of the parallels that Marx saw, his own historical parallel didn't quite work. The free-software system didn't get worse. "Over the course of the twenty-four months that I was watching it, the stuff kept getting better. The kernel got better. More drivers came out for this stuff. More people were using it."

This surprise prompted Young to talk with some key free-software users to find out why the system was such a success. One of his research subjects was Dr. Thomas Sterling, who worked at the Goddard Space Flight Center, just outside of Washington, D.C. In his conversation with Sterling, Young first glimpsed the wide variety of reasons for the success of free software. One of Sterling's employees, Don Becker, was writing Ethernet drivers that he licensed freely. Becker thought free software was "altruism" and thought himself part of the "altruism economy." But Sterling had a different view. As Young recounted the conversation to me, "Sterling said, 'Well, yeah, Don likes to think that way. But the reality is he's writing these drivers on Goddard time, and I'm the one who's writing his paycheck.'" In Sterling's view, the story was simple: this was part of a barter economy. "He was giving away something of relatively small value, and receiving back something of much greater value."

Young's a pragmatist. He's skeptical of accounts that rely upon a mysterious spirit. Free software came, he told me, not from a "community." "As far as I'm concerned, there's no such thing as a community. It's simply a bunch of people with a common interest." That "bunch of people" represented the "full range of humanity." But it had one thing in common: "a desire to see open-source software succeed." And that desire led members of this "bunch" to accept the idea of a commercial entity leveraging this sharing economy.

As Young became convinced that free software wasn't just a fad and, more important, that its success didn't depend upon reviving Lenin, he began to look for a way to build a Linux business. "I was looking for a product because I knew that given the growth of interest in Linux, it was going to end up in CompUSA.... I didn't want CompUSA as a competitor. I'd rather have them as a customer. So I was looking for products that I could get an exclusive on."

Young found a young entrepreneur to partner with named Marc Ewing. Ewing had for a time been developing a software tool to run on Linux. But after months of frustrating development, he concluded that what the world really needed was a better version of Linux. He therefore started to build that better version, which he would eventually name Red Hat Linux. Young heard about Ewing's software and contacted him. He offered to buy ninety days' supply of his beta, about three hundred copies. "There was dead silence at the other end of the phone," Young recounted to me. "I finally got out of Marc that he was only thinking of manufacturing three hundred copies. It was a match made in heaven." Red Hat Inc., was born, a paradigmatic example of what I call the hybrid.

Red Hat's success, in Young's view at least, came from something that seems so obvious in retrospect that it's puzzling more didn't try the same thing: that this free-software company actually made its software open-source. Other Linux distributions tried to mix open-source components with proprietary components. That was, for example, Caldera's strategy. But Young understood that the only way Red Hat could compete with Microsoft or Sun Microsystems was by giving its customers something more than what Microsoft or Sun could give them—namely, complete access to the code.

Young saw this point early on. He described a conversation with some engineers from Southwestern Bell at a conference at Duke. Young was surprised to learn that they were using Linux to run the central switching station for Southwestern Bell. He asked why. Their response, as Young recounts it, is quite revealing:

> Our problem is we have no choice. If we use Sun OS or NT and something goes wrong, we have to wait around for months for Sun or Microsoft to get around to fixing it for us. If we use Linux,

we get to fix it ourselves if it's truly urgent. And so we can fix it on
our schedule, not the schedule of some arbitrary supplier.

The key was to sell "benefits" and not "features." And here the benefit was a kind of access that no other dominant software company
could provide.

Red Hat is thus a "hybrid." Young was not in it to make the
world a better place, though knowing the man, I know he's quite
happy to make the world a better place. Young was in it for the
money. But the only way Red Hat was going to succeed was if
thousands continued to contribute—for free—to the development
of the GNU/Linux operating system. He and his company were
going to leverage value out of that system. But they would succeed
only if those voluntarily contributing to the underlying code continued to contribute.

One might well imagine that when a for-profit company like
Red Hat comes along and tries to leverage great value out of the
free work of the free-software movement, some might raise "the
justice question." Putting aside Marc Ewing (who had great coder
cred), who was Robert Young to make money out of Linux? Why
should the free-software coders continue to work for him (even if
only indirectly, since anyone else was free to take the work as well)?
What did the (in Young's mind at least) proto-Marxist Stallman
think about the exploitation of this work? "What about," one might
imagine the question being asked, "the ~~worker~~ coder?"

Yet what's most interesting about this period in the early life of
the hybrid is that there were bigger issues confronting the movement than whether a Canadian should be allowed to risk an
investment on a Linux start-up. The bigger issue was a general
recognition that free software would go nowhere unless companies

began to support it. Thus, while there was whining on the side-
lines, there was no campaign by the founders of key free software
to stop these emerging hybrids. So long as the work was not turned
proprietary—so long as the code remained "'free' in the sense of
freedom"[3]—neither Stallman nor Linus Torvalds was going to
object. This was the only way to make sure an ecology of free software could be supported. It was an effective way to spread free
software everywhere. And indeed, the freedom to make money
using the code was as much a "freedom" as anything was. If there
were people who objected strongly to this form of "exploitation,"
then, as Apache cofounder Brian Behlendorf described to me,
they "probably were disinclined from contributing to open source
in the first place. They might have kept themselves out of the
market and not spent their volunteer time or their hobbyist time
writing code."[4]

And thus Red Hat (and then LinuxForce [1995], CodeWeavers
[1996], TimeSys Corp. [1996], Linuxcare [1998], Mandriva [1998],
LinuxOne [1998], Bluepoint Linux Software Corp. [1999], Progeny
Linux Systems Inc. [1999], MontaVista Software [1999], Win4Lin
[2000], Linspire [2001], and Xandros [2001], to name a few) was
born. An ecology of commercial entities designed to leverage value
out of a sharing economy. The birth of the most important Internet
hybrids.

Now, as Red Hat demonstrates, there is a delicate balance to be
struck between the commercial entity and the sharing economy.
Red Hat succeeded in maintaining the loyalty of the community
because of how it behaved. It respected the terms of the license; it
supported development that others could build upon; indeed, as
Young estimates, at one point more than 50 percent of the core kernel development team worked for Red Hat,[5] and both Red Hat and

VA Linux Systems gave stock options to Linus Torvalds.[6] Many from the GNU/Linux community helped Red Hat understand what appropriate behavior was, and the company took great steps to make sure its behavior was appropriate. A key element to a successful hybrid is understanding the community and its norms. And the most successful in this class will be those that best leverage those norms by translating fidelity to the norms into hard work.

Perhaps the most interesting recent example of this model is a company called Canonical Ltd., a commercial entity supporting another brand of GNU/Linux called Ubuntu Linux. Launched in 2004 by the entrepreneur Mark Shuttleworth, Ubuntu aims to become "the most widely used Linux system." Its focus initially has been really really easy desktop distributions. (I've experimented with a number of Linux installations. This one was by far the easiest.) The company hopes that the ease and quality of its distribution (not to mention its price) will drive many more individual computer users to use Ubuntu Linux.

Canonical aims to profit from the community-driven and community-developed Ubuntu. Its vision is inspired by Shuttleworth, who says he has been "fascinated by this phenomenon of collaboration around a common digital good with strong revision control."[7] That collaboration is done through a community. Canonical intends to "differentiate ourselves by having the best community. Being the easiest to work with, being the group where sensible things happen first and happen fastest." "Community," Shuttleworth said to me, "is the absolute essence of what we do." "Thousands" now collaborate in the Canonical project.

To make this collaboration work, as Shuttleworth describes, at least three things must be true about the community. First, you must give the community "respect." Second, you must give "responsibility"—actually give the community the authority you claim it has. "If you're not willing to respect the fact that you've offered people the opportunity to get invested, and take a leadership position...there's no way that's going to grow a strong team."

Third, and maybe ultimately the most important: you have to "give people a sense of being part of something that has meaning." This the free-software community can give away easily. Contributors to this community "feel they're being part of something that's big and important and beautiful.... They feel like they get to focus on the things that they really want to focus on. And that's satisfying."

This is a common feature, Shuttleworth believes, across successful community-based projects. If you "look at Wikipedia," for example, "people genuinely feel like they're part of something: they're helping to build a repository of human knowledge, and that's an amazing thing. It's a full spectrum of motivation, just like you get the full spectrum of motivation in free software."

Shuttleworth's vision is different from Red Hat's. Remember, Young didn't believe in the "community" thing. Community is central to Ubuntu. But in this range of motivations, some tied to believing in something and some not, we can begin to get a sense of the interesting mix that the hybrid economy will produce. Diversity is its strength; it flourishes from the obscurity such diversity produces.

Beyond Free Software

Free software is the paradigm hybrid, in which commercial entities (Red Hat is just one) leverage value from a sharing economy.

But as with Wikipedia and the sharing economy, free software is obviously not the only hybrid. In this section, we will consider some other examples, and other flavors of this mix.

I've kept this list long because my aim ultimately is to convince you of the diversity and significance of this category of enterprise. But I've divided the examples into categories. Some hybrids build community spaces, some hybrids build collaborations, and some hybrids build communities. Consider each in turn.

Type 1: Community Spaces

From the very beginning of the Internet, its technologies have been used to build community spaces—virtual places where people interact, sharing information or interests. The people interacting do so for sharing-economy reasons: the terms under which they interact are commerce free, though the motivations for interacting may or may not tie to commerce.

Few have been able to translate these spaces into successful commercial ventures. Many are trying. The effort and the successes are examples of one kind of hybrid.

Dogster

Let's begin modestly with a hybrid that doesn't try to change the world, but has changed substantially how easily people can connect about their intense relationships (not to say "obsessions") with dogs. Dogster, as the Web site explains, is built by "dog freaks and computer geeks who wanted a canine sharing application that's truly gone to the dogs." Since its launch in January 2004, it has become the fastest-growing pet destination on the Internet, in 2007 "serving over 1.5 million photos for over 300,000 uploaded pets by 260,000 members; Dogster and Catster serve more than 17 million pages a month to over half a million visitors." The site offers "forums, classifieds, diaries, treats, private messaging, Gimme Some Paw, DogsterPlus, photo tagging, themed strolls, pet-friendly travel and pet-personality matrix." The site is designed to make this community space the dominant pet center on the Net.

Dogster doesn't do this for free. The community space supports itself through advertisement. Given the size of the community, the revenue is likely to be close to $275,000 a year.[8] But that revenue means "a couple of handfuls of people can be employed by that site." And because it can fund itself like this, the site "will touch a lot more people."[9] The site thus leverages the community of passion and conversation that surrounds pets to produce revenue that supports the site. A hybrid.

craigslist: "Like, Peace, Man"

The bread and butter of most local newspapers is advertising. The most lucrative of this advertising are classifieds. According to one estimate, "U.S. newspapers derive 37% of their total advertising income from classifieds."[10] For major papers, the number is even higher: "Revenues from classified ads account for around 43% of total advertising revenues from major papers and over a third of total revenues."[11]

In 1995, Craig Newmark launched a service that would change all that. craigslist was offered first as a community site just in San Francisco, enabling people to post free ads for everything from home rentals to offers for the erotic. The site grew. Fast. It incorporated in 1999 and then expanded into nine U.S. cities in 2000, four more in 2001 and 2002, another fourteen in 2003. By the end of 2006, there was a craigslist in more than four hundred cities around the world,[12] and overall page view growth was 195 percent that year. Today, it is the ninth-most-visited site in the United States and managed largely by Jim Buckmaster.[13]

But though the site has spread broadly, in critical ways the site has never really changed. Its design is extremely simple, almost retro now. There are no fancy graphics, no Flash introduction. When you navigate to craigslist, you're presented with a screen of blue text, each a link to a category holding the stuff that you might want. On the URL bar in your browser, the icon for the site is a peace sign.

Newmark originally launched his site "as a community service." Its success, at least so he believes, comes from the fact that "people can see that we actually do [provide a community service] and follow up on that."[14] The follow-up is in form as well as substance. Again, the simplicity of the site speaks volumes. Users are reminded of an earlier, noncommercial Internet. That simplicity stands in sharp contrast to the molasseslike sophistication of most of the rest of the commercial Web.

Second, "99 percent of the site's content is free."[15] This "free" reinforces the sense that people are exchanging information, or "bartering" information, in something other than a commercial economy. craigslist enables them to "share" information—wants or needs—as members of the craigslist community.

Third, craigslist reinforces that sense of community by shifting to the users a certain responsibility. The power to judge what content survives on craigslist is vested first in the community. As Newmark described it to me, "We're saying, 'Hey, if you see something that's wrong, you can flag it, and if other people agree with you, it's removed automatically.' People respond real positively to being trusted."

But fourth, and most interesting for understanding this hybrid, not everything on craigslist is free. While the site has clearly signaled—and said, again and again—that, as Craig Newmark told me, "we're not out to make lots of money," two classes of ads— ads for jobs in eleven cities and apartments in New York City—are not offered for free.[16] From this income, the balance of the site is supported, and its founders profit.

The community thus does not demand a commerce-free zone. It does not require that its founders remain poor. At least so long as the demand remains modest, the community stays. And it stays even though the revenue to craigslist is quite substantial—"estimated at more than $20 million per year, with very healthy margins for this tiny private company."[17]

Whether this "community site" really feels like a community is partially revealed by the sorts of things people think it is appropriate to do, or talk about, on craigslist. Some of those things I can't repeat in a book like this. But some plainly do signal something important about the sense people have of the craigslist community.

Take for example the site's response to the Katrina disaster. Immediately after Katrina hit, the users of the New Orleans craigslist took it over, in effect, and directed its attention to helping victims of Katrina cope with the disaster. As Newmark told me, "Survivors announced where they relocated to. Friends and family asked people, 'Hey, have you seen so and so?' And then later, well actually, pretty soon, people started offering housing to survivors. A couple days later, people started offering jobs to survivors." Three days after Katrina hit New Orleans,

craigslist's New Orleans page featured more than 2,500 offers from around the country for free housing for hurricane victims, ranging from "Start a new life in South Carolina" to "Comfy couch in spacious NYC apartment."...Never before has the Internet played such a vital role in filling the information void following a natural disaster.[18]

As the *San Francisco Chronicle* recounted:

> The message is short. So short it would fit on a postcard. It lingers in cyberspace waiting for a response. "Family of 4 willing, wanting to help. Can drive to get you. Stay as long as you need to here in Albuquerque. God bless you. We care. Howard and Lisa Neil." It is one of more than 2,000 classified advertisements—and counting—posted on craigslist, a network of online urban communities, offering free, temporary housing to people who have lost their homes to Hurricane Katrina.... The listing can be found under a new category on the Web site—Katrina Relief—that also includes listings for relief resources, missing people, temporary jobs, missing pets, transportation and volunteers.[19]

The point in recounting this story is not so much to praise craigslist (as if it needed more praise). It is instead to highlight what may already be obvious: craigslist's standing as a "sharing economy" was reflected in the fact that it was clear to everyone that that was the place to go to help survivors from Katrina. There were no doubt sites with a bigger presence. Wikipedia has more hits each day than craigslist. But the meaning of Wikipedia is not activism, it is knowledge. Yahoo! and Google both have a presence much bigger than craigslist or Wikipedia. But it would have been hard to generate the feeling that this was a community response by tying the activity to those commercial giants. And there's no need to waste time explaining why people didn't use the government's Web sites to work this good, so pathetic is the government as an inspiration for community. We see how the community understands craigslist by watching how the community uses craigslist. And when emergency help was needed, the obvious response was this simple site of community messages.

Of course, craigslist was not the only Internet response to Katrina. Neither was it the most important. David Geilhufe's PeopleFinder Project probably earns that title; it was built exclusively by volunteers in an extraordinary demonstration of a sharing economy that ultimately hosted more than 1 million Katrina-related searches in the "immediate aftermath of the hurricane."[20] But Geilhufe's success is not a complaint about craigslist. The significance of craigslist was that it was the place to start. Its karma made this commercial entity enough of a community site so that it made sense to use it to help the victims of America's most devastating natural (and then governmental) disaster.

It's impossible to say how long craigslist can keep this karma. Knowing Newmark, I'd bet forever. But institutions change. And sometimes, institutions change people. The importance for our purposes, however, is simply to suss out what makes the sharing salient in this commercial entity. Newmark's is the hybrid to envy. His intuition of how best to maintain this community is pure gold.

Flickr

In the early 2000s, Stewart Butterfield and Caterina Fake decided they wanted to build a multiplayer game called Game Neverending. They failed. Flickr was the result of their failure (would that we all could "fail" so well). Realizing the code they had built would make a great photo-sharing site, they launched the site in February 2004.

By September the site had over sixty thousand registered users. Six months later, that number topped four hundred thousand. In December 2006, five million users were registered at Flickr.[21]

Theirs wasn't the first site to enable people to post pictures to the Internet. In 1999, Lisa Gansky and Kamran Mohsenin started Ofoto, an online photography service. Ofoto was acquired by Kodak in 2001. Kodak spent millions building Ofoto. But Kodak's site was crassly commercial. Everything was about buying photographs, or buying albums, or buying T-shirts with your photographs on them. The site encouraged community in just the sense that a Kodak store at the mall encouraged community.

This failure at Ofoto wasn't for a lack of trying. I knew some of the team at that Berkeley start-up. They got it. They worked hard to make Ofoto what it should be. But what it should be was resisted by the powers at Kodak. Kodak didn't get community, even if its marketing department was good at producing sappy commercials that celebrated it.

Flickr was different. From day one, the aim was not to facilitate commerce. The aim was instead to build a community. People could easily share their photographs and get feedback from other photographers and friends. This is what distinguished the site from others. As Stewart Butterfield told me, at the time Flickr launched, "there really wasn't a concept of public photos."[22] Flickr changed that. In the beginning "80 percent of the photos [were] public." That meant that "there was a much bigger audience for the photos that were on Flickr."

One way Flickr signaled the freedom to share was by explicitly incorporating tags that enabled people to say, "You're free to share this work." Those tags came in the form of Creative Commons licenses. (We'll see more about Creative Commons in chapter 10.)

They signaled that Flickr's lack of control over intellectual property was explicit: the users owned the IP. They were free to license as they wished. Flickr was keen to encourage the idea that it licensed to enable people to share.

This focus on sharing helped build a certain kind of community. Flickr quickly became part of the identity of Flickr users. As Butterfield put it, "Netflix is an example of where I get the value out of other people's recommendations...but it's not part of my identity that I'm a Netflix user. [But] Flickr users actually have meet-ups in Tehran and Kuala Lumpur and Manchester." And community members do more than simply use the space that is provided for them. In a metaphorical sense, they pick up the trash. One of the keys to Flickr's success was the fact that its members constantly policed the site against porn. Members can flag a photo as inappropriate. Pornography is quickly moved off the site. The same with reviews. As Butterfield told me, "People aren't writing reviews just because they happen to like writing reviews." They do it instead because they feel part of a community.

In March 2005 Flickr was acquired by Yahoo!. Its founders continued to work for the company. Characterizing their job, Butterfield told me, "In some sense we're trustees or custodians...like a land trust that buys up wetlands." For obviously, Yahoo! didn't buy Flickr as a way to subsidize the sharing economy. Flickr instead was to be a model for the hybrid that Yahoo! intends to be. Yahoo! intends to profit from this community of photo collaboration. So far, however, the company has been modest in its aspirations. Most revenue to Flickr comes from Flickr memberships—which give users "unlimited storage, unlimited uploads, unlimited bandwidth, unlimited sets, personal archiving of high-resolution original images, ad-free browsing and sharing." Some revenue comes from partnerships with sites that offer prints of Flickr photos. But so far the company leaves millions on the table. Butterfield recognizes this: "We have well over a billion page views a month now. That's one of the biggest sites on the Internet. And if we just went for the maximum amount of graphical advertising…we'd make a lot more money." But, as apparently Yahoo! also recognizes, "it just wouldn't last for very long." Respecting the norms it understands its community to carry, Yahoo! thus continues to let the community live like a sharing economy, with small but increasingly significant efforts to leverage something on top.

YouTube

In 2005, three early PayPal employees—Chad Hurley, Steve Chen, and Jawed Karim—started building a video-sharing service for the Internet. They weren't the first. But they architected the best. Steve Chen told me, "The sort of initial acceleration for our growth came from the technology. We did some things right—namely, choosing Flash video as a delivery platform so you didn't have to download anything. The video just plays in the browser."[23] By using Flash as the video format, they guaranteed that anyone with a (modern) Web browser could view the videos. (As Chen told me: "I always thought about the grandmother in the Midwest, if she came to a video site.") Very quickly the service grew. Indeed, by the summer of 2006, YouTube was the world's fastest-growing Web site. Nielsen/NetRatings estimated traffic was growing 75 percent per week in July 2006. One hundred million clips were viewed daily; sixty-five thousand were uploaded every hour. Users spent twenty-eight minutes per visit on the site. In the United Kingdom the site quickly became the largest online video market.[24] In 2007, YouTube was bought by Google—for a reported $1.65 billion.

So this success came first from great code. But technology was not

everything. The balance was, as Chen put it to me, "the community...the people's relationship, their tie to, user-generated content on
YouTube." This value came directly from the community. YouTube
users select the content to be added. They make the content that gets
added. Some of YouTube's content is copyrighted material that the
copyright owner didn't upload. But if the top hundred videos is any
indication, most of the most popular of YouTube's content comes
from users creating content that they then upload to YouTube's site.
The site has become a bizarre mix of the most bizarre video content. It has launched some stars, and some fanatics. (Wikipedia lists
more than sixty YouTube creators who have become Internet phe-
nomena on the basis of their appearance in YouTube videos.)[25] No
site—ever—has more quickly become central to popular culture.

So why do people do it? What do they expect to gain from
working so hard to make a couple of Stanford dropouts rich? Most,
following Dan Bricklin's insight, contribute as a by-product to getting what they want—a simple, cheap, and effective way to spread
their video. YouTube, and the other video-sharing sites, provide a
service that even three years ago seemed unimaginably difficult: a
network location that would deliver anyone's video for free.

But some do much more than simply consume content. As
with craigslist, the community of YouTube users helps police the
YouTube content. Inappropriate content gets flagged. Content violating the rules gets reported. Like neighbors in a well-kept community, the users clean up after one another and take pride in the
place they've helped build. The result is a space that is addictive as
well as amazing. Its biggest draw, founder Chen told me, is

just the content itself. We see on our top-viewed pages some of this
content from [professional sites] versus some of this content from

user generated. They're just married together, sitting side by side on all these "top browsed" lists. But some of this stuff was created with $500 of editing equipment and a lot of time spent on it versus, on the other end of the spectrum, probably millions of dollars to create this fifteen-second commercial.

"The very nature of UGC [user-generated content] video sites, as well as their potential for financial success and sustainability, relies on the effective leveraging of Internet-based social networking activities. In other words, the content must be shared in order to represent value."[26]

TYPE 2: COLLABORATION SPACES

A collaboration space is different from a community space like Flickr or YouTube. Those participating in a collaboration space think their work is different. Or more accurately, at least some (significant portion) of those on a collaboration space believe they are there to build something together. The community is visible. It's the focus of the work. And the product of the participation is intended to be more valuable than the material they found when they came.

This collaboration can come in many forms. Consider a range of examples.

Declan

Declan McCullagh is a journalist. He began his career covering and protesting Internet censorship. In 1994, while a student at Carnegie Mellon in Pittsburgh, McCullagh began organizing against efforts by CMU to "remove any Usenet newsgroup that had the word 'sex' or 'erotica' in the title."[27] That organizing took the form of an e-mail mailing list, Fight Censorship. The mailing list had two channels. One was an announce list, which McCullagh published to, posting information about the fight against censorship. The second was a discussion list, on which individuals receiving the e-mail could also post comments or replies to what they received. The traffic from the second list was often overwhelming. That led most to stay on the announce list only.

Fight Censorship was soon renamed Politech, and its focus broadened to include the "growing intersection of law, culture, technology, politics, and law."[28] In the terms that I've used, Politech is a sharing economy. At least those in discussion space speak out and/or listen, constituting a community of common interest and, sometimes, common action. (I criticized [mildly] McCullagh in my first book; the flames I got from McCullagh's community were anything but mild.) The community in this sense is like hundreds of thousands that live in discussion spaces everywhere. Yahoo! has over 2,083,698 such groups.[29] So too does Google. McCullagh's is much smaller than those, but for a private (noncommercial) site, it is quite impressive: the list started with a couple hundred readers. It now boasts more than ten thousand subscribers.

But McCullagh has transformed this sharing economy into a hybrid. For in the time since CMU, he has become a professional journalist. His journalism is within the scope of the interests of Politech. Neither aggressively nor inappropriately, McCullagh uses the community he has built to better his journalism. As he explained to me: "Intentionally, I tried to create a community. I never really thought much about where it would end up. But I did

do things like that [from the start]." The reasons are not hard for a
journalist to identify with. Said McCullagh,

> This is a problem for journalists…because people read our arti-
> cles but we don't really have a community to talk to. We just get
> nastygrams from people who hate us, and we get nice notes from
> people who like us. But the process of being able to develop an
> article idea with some community input is very valuable. So you
> can throw out sort of half-baked ideas on a mailing list, and then
> hone those into something that you can get paid to write a few
> days later.

When he was hired recently by CNET, he and his new employer
agreed that he would get to keep his community and to feed it in
the way its members had come to expect. Nothing confidential
would be shared, but the community could be drawn upon to help
this journalist produce good work.

And not just good work. Politech members gather for dinners
in big cities. McCullagh connects with them as he travels. The site
thus spawned a community—and not just one community. As
McCullagh told me as we ended our interview, "Oh, I should say: I
met my wife through the list."

Slashdot

Slashdot began in September 1997 as a Web site covering technology-
related news. Its angle, however, was a technology that enabled
users to both comment upon the articles that got referenced, and
comment upon the comments. The consequence of the second set
of comments would be to filter out comments not thought useful.
That meant the site could self-edit, and hence present to any reader

a high-quality public debate about issues important to the technology community.

Today there are more than a quarter million people collaborating in just this way on Slashdot.[30] Their work—work they don't get paid for, don't even get frequent-flier miles for—produced a site worth millions of dollars. In 1999, Slashdot was sold to Andover .net, and ads were added to the layout. So readers edit, and readers and Andover.net then profit.

Collaboration here winnows a potentially endless array of comments to a relative few that people reading the site would or should want to see. The site adds a kind of collaborative editing to a news page that is RW to the tradition of RO news. This collaboration produces a high-quality RW site. Editing is value. This value is produced for free.

Last.fm

No doubt the industry that has squealed the most (think stuck pig) about the Internet is the traditional recording industry. As I've described, the Net competed (whether legally or not) with its model for profiting from music. It fought hard to limit that competition. But not all in the music business have fought the Net. Some have tried to build upon its design, to enable collaboration among music fans. One great example of this technique is the site (recently acquired by CBS) called Last.fm.

Last.fm's objective was to find a way to map user preferences for music into an engine that might recommend music more intelligently. Many have tried this. But Last.fm does it by a unique mix of community and technology. The technology watches what you listen to, "scrobbling" (meaning sending the name of the song you're listening to) to an engine that then learns more about you (and people like you). That engine then enables individuals to be linked with others. But not just anonymously. Instead, the technology helps individuals link their own user pages with others, both friends and those within their musical "neighborhood."

As Japanese venture capitalist (and Last.fm investor) Joi Ito described it to me,

> The Last.fm community originally was, and may still be, revolving around the cleaning up of the data. So if you have titles of songs that are misspelled, or if you have an artist whose name is written differently in Japanese, there's a whole community of people who go in and fix those ambiguities and fix the data.

The contributions, however, are now much wider than some form of community editing. "There's also a community around each of the bands where people talk.... There are discussion groups and people are contributing information. And by listening to a lot of the music, the users are creating profiles."

Thus, simply by listening to music, members "generate value for the community." Listening becomes a kind of advertising. Each song you listen to gets tagged as a song you listened to. It advertises the song. It cues others to your interest. But this advertising is simply conversation. Again, Ito: "What we're currently doing in terms of so-called advertising is really part of the conversation." Volunteers converse. The product is a value to the company.

Microsoft

In an increasingly remote region of cyberspace called Usenet, a fanatically committed group of volunteers works to help people they've never met with computer problems. These problems might be simple. Some are quite complex. Yet these volunteers spend many hours helping these lost cybersouls find digital salvation. In the particular space to which I am referring, there are over 2 million contributors a year, with over forty thousand making more than thirty-six contributions each per year, and about eight hundred making contributions just about all the time.

The striking thing about this story is not that there are people helping other people. Nor is it that the people are helping people they're never going to meet. It is instead that all this pro bono effort is devoted to a very peculiar end: that of making Microsoft's customers happier. For these volunteers live within Microsoft support "newsgroups." They're not paid by Microsoft. The vast majority are not even recognized by Microsoft. But they all work (and some work quite hard) to make Microsoft richer by solving its customers' problems.

Microsoft knows this. In one of two buildings devoted to research, the chief of Microsoft's Community Technologies Group, Marc Smith, leads a team carefully studying the behavior within these groups. The company has developed elaborate technologies to measure the "health" of these and other virtual communities, asking, for example: Is there enough balance to the contributions? and Are the contributors constructive or smothering? It is constantly reflecting upon the husbandry necessary to make these communities work better. Researchers study the interaction. They watch the pattern of communication. They try to learn what forms of interaction work.

The problem isn't easy. There are scores of variables to consider. But the one variable that remains completely absent is money. Not because Microsoft is too poor to pay its help. Or because the community is not creating important value for Microsoft. Money is absent because Smith doesn't believe that "cash exchange for community support" helps. "There are numerous social benefits that amply incentivize contributions to communities," Smith explains. Money isn't one. Indeed, for the reasons we've discussed, money would probably hurt.

This is Microsoft building a hybrid economy. Volunteers living within what was once a pure sharing economy—Usenet—devote extraordinary time and effort to helping Microsoft users better use Microsoft products. Microsoft is not passive about this sharing. It cultivates it. It spends real resources to understand how to make it work for it, better. But the product is a network-based community collaborating to make it easier to use Microsoft products.

Yahoo! Answers

Usenet is not the only forum for sharing information. Increasingly companies are building "answer sites" to encourage a wide range of questions that will encourage a rich and diverse community. In December 2005 Yahoo! followed the Usenet lead and launched a service called Yahoo! Answers.

The basic idea was simple enough: Millions of volunteers would spend their free time answering other people's questions, for free. If millions did this long enough, and reliably enough, Yahoo! would profit from it. Yahoo!'s aim was to become the hub of community-based activity on the Web. The hottest answer site would be an important part of that hub. As described to me by Yahoo! founder Jerry Yang,

> We've been at it for about a year and there are 75 million people a month participating either asking questions, answering questions, looking up answers. You know, some of it is silly, "Why is the sky blue?" Others are very technical questions or very specialized questions around taxes or profession. A lot of people saying,

"I can't get my Mac to work. What do I do?" And all of a sudden, you have this very natural human interaction of "I have a question and somebody out there must have an answer for this." But you bring it on the Web and then you have a global community.[31]

As of April 14, 2008, the site contains 35,411,866 questions and 35,411,851 answers.[32]

Again, why would anybody want to help Yahoo! like this? We've seen the answer before. They don't do it to help Yahoo!. Some like to show others how smart they are. Some like to help. "Because they like to" is explanation enough for why people offer their expertise.

Yahoo! adds to these incentives, not with money, but by offering a point system to users. Users get one hundred points when they open an account. They get two points for every answer given, one point for every vote on an unresolved question, and ten points if their answer is selected as best. To ask a question costs five points. If you delete an answer you posted, you lose the two points you got by posting it, and you lose ten points if a question or answer is deemed to violate the terms of service.[33]

But, as with Microsoft, money is not part of the equation. The incentives from a gamelike structure are enough; adding money would make this seem more like work. For the users, it isn't meant to be work. Again, Yang: "I'm not sure we'll just do, 'Hey, you know what, if you answer a lot of questions, you'll get paid.' I think there is this fine balance between paying people versus people feeling like there's a noncommercial motive."

Wikia
Wikipedia, as I described in the previous chapter, is a sharing economy. It has sworn off advertising as a way to raise money. This is no

idle resolve: as I've already mentioned, based on the traffic Wikipedia garners, it could earn over $100 million a year if it added advertising to its site. Such is the opportunity of top ten Internet portals.

Wikia, another wiki site, was launched by Wikipedia founder Jimmy Wales. Its aim is not to build an encyclopedia. Rather, its aim is to be a "platform for developing and hosting community-based wikis. Specifically, Wikia enables groups to share information, news, stories, media and opinions that fall outside the scope of an encyclopedia.... Wikia is committed to openness, inviting anyone to contribute web content."[34] The site is enjoying the Jimmy Wales magic. With eight hundred thousand articles, it is actually growing faster than Wikipedia was at a comparable period.[35]

The site is already a treasury of human culture. Fans of television shows detail facts about the shows. The Marvel Database Project hopes to become "the largest, most reliable, and most up-to-date encyclopedia about everything related to the Marvel/DC universe." Football (meaning soccer) fans can build upon the "football wikia" for all topics related to football. Foodies can contribute to discussions of food and recipes. All of this work—or better, this passion for these subjects—is offered for free. No one makes money on Wikia.

Except, of course, Wikia Inc. For unlike Wikipedia, Wikia does run ads. Unobtrusive, sometimes silly, but increasingly valuable for the site. Wikia offers its users a free platform to build a community. They do the building. That building is a complex process of collaboration. Wikia gets the advertising revenue.

This fact leads many to wonder how such a site can work. Don't the builders of these wikis realize that Wikia could grow rich off of their creative work? The answer is yes, they do, but that doesn't stop them from contributing. Just as a bowling alley does, Jimmy

Wales explains, Wikia provides a context in which people get to do what they want. Like a bowling alley, people are happy if they get to do something they enjoy. No one begrudges the owner of a bowling alley his profit. Wales believes no one will begrudge Wikia its profit.

This is true, at least if certain other conditions remain true. There's got to be competition among wiki sites that allows users to move as they want. And Wikia supports this competition by enabling users to move the content of the wiki elsewhere if they begin to find the bowling alley no longer reflects their values. Wales imagines a Wiki user reasoning:

> "Look, we accept that you have ads because we know we need the infrastructure, but that's not giving you carte blanche to basically slap way too much stuff all over the site." And so that tension and the fact that communities are empowered to actually leave—they can take all their content and leave if we're not making them happy—are really important.

There's thus a social contract between the commercial and the sharing. Upon that contract, the value in Wikia gets built.

Hybrid Hollywood

Perhaps the most interesting emerging (if slowly) example of collaborative hybrids comes from the least likely of sources: Hollywood. Increasingly, Hollywood is including the audience in the process of building, spreading, and remaking its product. That practice constitutes a kind of hybrid.

The story is not always pretty. Progress has not always been smooth. Perhaps the best example of this struggle, producing in the end real understanding, begins with a home-schooled fourteen-year-old named Heather Lawver.

In January 2000 Lawver started an online newspaper called *The Daily Prophet*. This was not a religious paper. It was instead an effort to explain and extend the story given to her generation by the extraordinary author J. K. Rowling. Every day for months, Heather would collect articles written by kids around the world about the Harry Potter saga. She would edit them and then publish them online.

As Henry Jenkins, perhaps the world's leading scholar on the emerging RW creativity of the Net, notes, "Rowling and Scholastic, her publisher, had initially signaled their support for fan writers, stressing that storytelling encouraged kids to expand their imaginations and empowered them to find their voices as writers."[36] In 2003 Rowling welcomed "the huge interest that her fans have in the series and the fact that it has led them to try their hand at writing."[37]

But the instincts of a brilliant writer are not always the instincts of a media company. As Rowling's success migrated from the printed page to a major Hollywood media company, Warner, the "control" over what was now Warner's "property" shifted from a storyteller to a pack of lawyers. As Marc Brandon, the twenty-something Net-head Warner eventually brought in to deal with the problem they were about to create, explained, "Warner Bros. at the time... had never dealt with something the scale of a Harry Potter. Certainly not online."[38] So the company dealt with the Harry Potter questions in the way they had dealt with IP issues on the Internet. As Jenkins describes: "The studio had a long-standing practice of seeking out Web sites whose domain names used copyrighted or trademarked phrases.... Warner felt it had a legal obligation to police sites that emerged around their properties."[39]

"Police" in this context means firing off angry letters written by entry-level lawyers who always wanted a gun, but instead were issued IBM Thinkpads.

Lawver learned of these threats in December 2000. They transformed her into an activist. (Why? I asked her. "I think it just kind of came from a common sense point of view, and, also...I grew up in a household with three brothers, and they were all Weird Al fans. And so I was really familiar with his various battles against other artists.")[40] Two months later, she had organized a boycott of Harry Potter products. On February 22, 2001, the "Potter Wars" began.[41]

Lawver was the commander in chief. The teen debated Warner Bros. representatives on MSNBC's *Hardball with Chris Matthews*.[42] Newspapers around the world started picking up on the fight.[43] "We weren't disorganized little kids anymore. We had a public following and we had a petition with 1500 signatures in a matter of two weeks. They finally had to negotiate with us."[44]

Lawver's campaign of course leveraged the Net. Warner quickly became savvy in its strategy of intimidation. It avoided threatening Lawver directly; it hoped to avoid her following generally. But, as she told Jenkins,

They attacked a whole bunch of kids in Poland.... They went after the 12 and 15 year olds with the rinky-dink sites. [But] they underestimated how interconnected our fandom was. They underestimated the fact that we knew those kids in Poland and we knew the rinky-dink sites and we cared about them.[45]

"If someone got threatened by Warner," she explained to me, "they could come to us. And it got to the point where Warner Bros. was so afraid of me and my partner, Alastair Alexander, that if we sent them an e-mail, most of the time the threatening would stop." And "the most important" part of the story, "regardless of any legal impact it had," was that after this battle, kids from around the world were fighting back. They were "fighting their own battles now, because they have the confidence to do what they can."

This was the part of the story that I had heard about. It was the part, in my perverse yellow-journalism sense, I wanted Lawver to tell me more about. But to my surprise, and (eventual) delight, Lawver was not so interested in trashing Warner. Her real interest was in making me understand a different, less-reported part of the story.

For this was not simply a story of the big bad media company. It was also a story of a company coming to learn something about the digital age. As much as she was rightly proud of the movement that she had spearheaded, Lawver was also proud of the way she had helped Warner understand the twenty-first century. "We did a lot of educating," she told me, "regarding where fans are going to take a stand and how much crap we're going to take before we fight back." More important, she pushed Warner to understand that fans were not a burden. "Warner Bros. [came to]...realize that, 'Hey, these people are funding our franchise with their pocket money. We shouldn't scare the living daylights out of them.'" Fans were, she explained to me, "a part of your marketing budget that you don't have to pay for."

Lawver's story was confirmed by Warner Bros. Marc Brandon, now a vice president at Warner, was the contact with whom Lawver worked. He saw the history much as she did. "There was a 'better safe than sorry' approach" taken at the start, Brandon told me.

"Why don't we do as we've always done with our trademarks, and pursue as much protection as is humanly possible?" The answer to this question was Lawver: she showed them why. And Brandon was quick to admit that "some mistakes were made."

Very quickly, Brandon pushed Warner to "approach this from a practical point of view." He said, "There's a legal analysis that happens, but that was less of the discussion that occurred when this came up. It was more about how we can allow the fans to enjoy Warner Bros. properties online while still continu[ing] on with our interest as a company and a content creator."

The studio thus "went through an interesting learning process" about "how to interact productively with the fans." The "responses from the fan community," Brandon explained, "drove [Warner] to reexamine our approach. The process there was not, 'Well, let's sit down and look at the law and figure out what we need to do.' It was really more of a 'how do we deal with this practically?'"

"Practically" meant deciding which kinds of uses Warner needed to control, and which kinds not. And that line was not strictly a commercial/noncommercial distinction. Porn and child exploitation were obviously out, whether commercial or not. But there was more. "Our main concern was taking the properties of the company and utilizing them for direct commercial activities, such as creating consumer products based on [Harry Potter properties].... But there are other types of commercial uses that...we decided we could live with.... Some of those included banner advertising and affiliate programs."

This was Lawver's understanding of a fair deal as well:

Well, obviously, the line includes whether or not you're trying to make money off of the franchise that someone else owns. That's

the most important thing, because I do still respect Warner Bros.'
right to say, "Hey, you can't make money off of something that
we created and we own the rights to." So, that's one of the biggest
things we always tell the kids is that you can't start a Harry Potter
site and start confusing the fans and trying to get them to send
you money. Making money off of it is an absolute no-no.

But part of Lawver's message to Warner was that the fans were
where the money was. The consequence of (what the lawyers saw
as) this "piracy" was, paradoxically, more money to the "victim."
So while "making money off of [Warner]" was "an absolute no-no"
for the fans, it was not a no-no for Warner. Indeed, it was the argument advanced to Warner for why they should just lighten up.

Warner's Brandon was not eager to frame it in just that way to
me. He believed, as Lawver did, that "Warner Bros. has come to a
place that is very positive and mutually beneficial for, not only the
company, but the fan base as well." When I pushed him about the
benefit that Warner got, however—when I characterized it as more
profit to the company—he hesitated.

I think that, as a by-product of that, you certainly get to a benefit
to both the fan and the company, and I think that that's what's
important to Warner Bros. It's not solely a commercial interest.
It's not an approach of, "Well, let's use the fan so that we can make
sure and make a lot of money." That's not the number one goal of
the company.

I let it go. But the hesitation was revealing. What else is Warner
supposed to do, except "make a lot of money"? Sure, it must do that in

a way that doesn't offend its authors or its audience. But that's to say, it should "make a lot of money" over the long run. The unwillingness even to acknowledge what at some level must be true reveals either a careful marketing sense or a discomfort with the future. Brandon had helped build a media hybrid. He hadn't yet come to terms with just what that would mean for the company.

Warner had learned that being less restrictive with its intellectual property strengthened fans' loyalty to the brand and, hence, the return to its artists. Again, if that is true, then of course Warner should be less restrictive with its IP. And we all should be happy about this because (a) Warner's artists succeed more, and (b) IP rights will restrict less. Or more precisely, because (b) IP rights are not restricting in ways that limit the freedom of kids without any benefit to Warner artists. More freedom is good, especially when less freedom would help no one.

People still resist this idea, however, by saying that there's something unseemly in Warner "exploiting" kids. But I don't see it that way. Warner's "exploitation" consists in giving kids more freedom than they otherwise would have, under the rights that the law legitimately grants to Warner. Perhaps you have quibbles with how broadly the law grants Warner rights. That's a fair argument for Congress, but not a fair complaint against Warner. Warner is releasing rights given to it. Because of my (perhaps retrograde) views about corporations, I hope it is doing so because it helps it make money. And because of my (perhaps retrograde) views about transparency, I would hope we could praise it for doing what it does for the real reason it does it.

With this freedom (granted to the fan) comes a loss of control (by the company). Freedom will inspire a wider exposure of the

original content. But as Jenkins puts it, "it also threaten[s] the producer's ability to control public response."[46]

Tapscott and Williams make the point more generally:

> So here's the prosumption [meaning PROfessional + conSUMER]
> dilemma: A company that gives its customers free reign to hack
> risks cannibalizing its business model and losing control of its
> platform. A company that fights its users soils its reputation and
> shuts out a potentially valuable source of innovation.[47]

Yet at the same time, media companies, as well as businesses
more generally, are recognizing that closer ties with the audience are
the key to profit in the era of the hybrid. As Jenkins writes, "Rather
than talking about media producers and consumers as occupying
separate roles, we might now see them as participants who interact with each other according to a new set of rules that none of us
fully understands."[48] Artists who get this are building "a more collaborative relationship with their consumers."[49] And "in the future,"
Jenkins argues, quoting cultural anthropologist Grant McCracken,
"media producers must accommodate consumer demands to participate or they will run the risk of losing the most active and passionate
consumers to some other media interest that is more tolerant."[50]

Harry Potter, of course, is not the only example of Hollywood
leaving the twentieth century. Practically every major franchise of
content is coming to understand the value of the community of fans
who work (for free) to promote their content. My favorite example
is the fandom surrounding the series *Lost*. *Lost* is currently in its
fourth season, but it has already become "a model for a new media
age," as the *Los Angeles Times* put it.[51] The show has encouraged a
wide range of unregulated fan sites—including the wiki Lostpedia

and many more sites on Wikia, as well as twenty other platforms whose content can be accessed and shared. These sites help viewers understand and keep up with the show. They give fans a place to explore and add to their obsessive understanding of the detail of the show. And by being free with the content, *Lost* producers Damon Lindelof and Carlton Cuse can avoid their "biggest worry": "losing viewers who skip one episode and don't return because they become intimidated by the revealing flashback they missed or the complex plot's twists and turns."[52]

Thus, ABC is using a sharing economy to drive interest and eyeballs to commercial product. How long this model can be sustained no one knows. As Lindelof said: "We're exploring a new frontier here.... So it's best to see what it is first, as opposed to everybody walking up to the cash register and saying, 'Pay me, and then we'll do the exploring.'"[53]

TYPE 3: COMMUNITIES

I've described community spaces and collaboration spaces, intending the word "spaces" to reduce a bit the grandness of these other, big words. One might call Dogster a community. In a sense it is. But the range of life lived in Dogster is much narrower (one hopes) than the range of life in a typical community.

Yet there are spaces on the Net that aspire to be more than community spaces. And some of these spaces do deserve the moniker "community" without any qualifiers. For old-timers, perhaps the first (and maybe the best) such space was a community called the Well. But for the generation that knows the Net right now, the most interesting examples of hybrid communities are virtual spaces like Second Life.

Second Life

Think of Second Life as a world that a character you create gets to travel to. Once in that world, the character gets to do just about anything. The character (or, forgive me, but better, "you") can hang out with others. Buy some designer clothes. Create a new kind of motorcycle. Buy some land to build a house. Grow flowers. Basically anything you can do in real life, you can do in Second Life, so long as you'll let the word "virtually" attach to much of what you do. You can have a long conversation with someone you met in a virtual hot tub. You can explore the extraordinary virtual worlds that others have built. You can build an extraordinary virtual world that others can explore.

Of course, there's nothing about virtual worlds that makes people more virtuous. So as well as all these things that we'd be proud to tell our parents about (or happy our kids did), there is lots that happens in Second Life that we'd never tell anyone about, or that we might wish our kids would never do. Sex is the organizing term for this set of stuff. Second Life has lots of virtual sex. Some you might have no problem with—consenting virtual adults, and all that. Some you might have lots of trouble with—consenting virtual adulterers. But some you might wonder whether you should have any problem with: Would you be happier if your son was experimenting with real space? Would it be better if your roaming spouse was meeting someone at a bathhouse? Sure, the virtual both creates and satisfies demand. Maybe without Second Life, your spouse wouldn't wander at all.

People doing what they want to do doesn't on its own create a sharing economy, any more than the passengers on an Airbus 310 constitute a sharing economy simply because they all happen to want to fly to Canada. Second Life, like Vegas, gives people a chance to do something they want to do. And people spend hundreds of hours doing that. But much of the stuff that members of Second Life do is stuff that builds value for Second Life. Linden Lab, the creator of Second Life, has built a " 'product' that invites and enables customers to collaborate and add value on a massive scale."[54] The members do this with one another (mainly) for free; the product of what they do for free is a much richer, more interesting virtual world for Linden Lab to sell membership to.

We can measure the community here by mapping the range of public goods (meaning goods which everyone in the community gets to share) that community members create.

First, Second Lifers contribute the good of help. A significant percentage of Second Lifers hang about, trying to help newbies as they learn the ropes. As the chief Second Lifer (i.e., the Boss), Philip Rosedale, described to me,

> Second Life is a daunting environment because, obviously, it's so rich with capability. And, so, as a new user to it, you are bound to be frightened, confused by, or frustrated in your attempts to do pretty much anything that you're trying to do. And, I think what has come of that has been quite interesting.[55]

For, he continued:

> Once you have figured out how Second Life really works, what tends to happen is you then have a kind of currency that you can freely give, which is your advice and explanation of how things work. And, I think that that behavior is so prevalent. It's so pleasurable to tell a new person what to do, and they get so much from it, as well, and are so thankful and then you just have that natural, I don't know...I think there's a natural pleasure that we take as humans in being able to help

somebody else that, I would say, that the general idea of helping each other in the environment is a very strong one, and you see, nowadays, you see numerous cases of that happening in a way that doesn't...isn't really, I suppose, economically rational at a tactical scale.

So does that mean Second Life was constructed to create this kind of community? "No," Rosedale said with some resignation. "I don't think we've explicitly tried to make things hard in Second Life. I think we managed to accidentally do that pretty well."

Second, Second Lifers contribute the gift of beauty. Like people in a well-kept neighborhood, Second Lifers spend endless hours making their property beautiful. Not just to make it salable, but also to make the neighborhood more attractive. They build new designs; they add painted posters or images; they craft gardens, or parks where people can meet.

Third, Second Lifers contribute code. According to the company, some 15 percent write "scripts" in Second Life, meaning the code that builds things or places that others can see. A significant portion of those code writers—at least 30 percent—make their code free for others to share.

Fourth, Second Lifers have built institutions that make Second Life work better. One member, for example, Zarf Vantongerloo, decided Second Life needed a way to authenticate statements or promises (as in contracts). Using encryption technology, Zarf built a Notary ("Nota Bene" in a land called "Thyris"). When a person signs the document, the Nota Bene adds a cryptographic signature which verifies both that the signature of the parties is genuine and that the text on the document hasn't been changed since it was signed. The program is open-source, and uses open-source encryption routines (providing a way for people to trust that the program does what it says it does). The code thus adds a bit of confidence into the system, one "baby step" to better governance, as Zarf put it.[56]

Finally, Second Lifers struggle through acts of self-governance. The city of Neualtenburg (literally, New-Old-City) is an example of this. Neualtenburg describes itself as a "one-of-a-kind self-governed community, whose purpose is to (1) enable group ownership of high-quality public and private land; (2) create a themed yet expressive community of one-of-a-kind builds; and (3) implement democratic forms of self government within Second Life."[57]

The city builds this community through a mix of architecture, culture, law, and politics. Unlike most spaces in Second Life, the design of Neualtenburg resembles a medieval Bavarian city—wandering, organic, nonrectangular. The city is designed to be a "nexus for progressive social experimentation...including modern art...political organizations...and education." And the city was the first "democratic republic" built within Second Life. It has a nonprofit land cooperative; it has offered the first "well-defined investment bonds"; and it has Second Life's first and only constitution. In addition, the government offers well-defined and binding land deeds and covenants, and a virtual-world legal system.

These are all the sorts of things that members of any community do. They all create a kind of value that more than the creator gets to share. And as with any community, the more people contribute, and see others contribute, the richer everyone feels. Indeed, for many, the richness of the life they know in Second Life exceeds the richness of the life they can live in real life. As Sherry Turkle reported from one user of a different but related world more than a decade ago,

I split my mind. I'm getting better at it. I can see myself as being two or three or more. And I just turn on one part of my mind and then another when I go from window to window. I'm in some kind of argument in one window and trying to come on to a girl in a MUD in another, and another window might be running a spreadsheet program or some other technical thing for school. . . . And then I'll get a real-time message [that flashes on the screen as soon as it is sent from another system user], and I guess that's RL [real life]. It's just one more window . . . and it's not usually my best one.[58]

Second Life is of course not the only virtual place in which people build this sort of community. Many (perhaps most) of the most interesting virtual games have this component built in. Japanese entrepreneur and venture capitalist (and gamer) Joi Ito describes the community in perhaps the Net's most popular game, World of Warcraft: "[T]he game emphasizes the necessity for a group. As an individual, you realize that the only way that you could create a group is by sharing and by being friendly." This is a lesson especially important for kids.

So a lot of times when young kids start playing World of Warcraft, in the beginning they're very greedy. They look at it as a game and they don't look at these other people inside of this game as real people. They will do something stupid in the middle of a raid, or they'll go off and leave the game without saying anything, and do something that will cause him to be mean, so . . . they realize that they quickly lose their community. Once you lose your community, you're unable to then go off and even set foot inside of certain dungeons and things like that without the right kind of . . . without a group.

The game thus "pushes you to help each other. . . . It is very difficult to play by yourself, and being nice to people and gaining friendship is rewarded." The design thickens the community as a way to improve the play of the game.

Around these sharing economies, companies like Second Life build business. They thus try to extract profits from the sharing of others. And again, in my (what to some seems) Neanderthal view, this search for profits is to be praised. Of course, everyone should understand just what's happening. Transparency is key. But assuming this understanding, the success of the company means a greater opportunity for community by the members. There may be squabbles about the terms: Should it be cheaper? Should the company make more? But the structure of the deal is not a mystery. Linden Lab will, if successful, make money by giving its customers just what they want: more than just a service, an experience that gives at least some of the rewards of any sharing economy.

So far, the description of this hybrid has been at the content layer of Second Life. In 2007 Linden Lab took one further step down the hybrid path, by "open-sourcing" the client with which one interacts with Second Life. The code for that client was placed under the Free Software Foundation's General Public License (GPL), meaning the source was free for anyone to tinker with, and modifications of the source when distributed had to be distributed openly as well.

Rosedale's motivations for this change are mixed. One part is the belief in his company, and the belief that it should grow as fast as possible. Second Life makes people better; Second Life should therefore grow as quickly as possible.

> And, it just seemed that, if your mission as a company is to grow as
> quickly as you can, which very much is our mission, not just from
> the "make money" perspective but actually, I think, primarily from

the perspective, which I've talked about in our blog quite a bit, we broadly believe that this is a force for good. That as people use Second Life it is a statistical improver in one's quality of life. And, if that's true, then we should allow it to grow as quickly as people want it to, and we should criticize ourselves if there's a moment at which we maximize revenue at the cost of growth or maintain some degree of control that, if we had given it up, would have caused this thing to grow faster. So, we made the decision to open-source.

Second, the decision was driven by the recognition of how much value Linden Lab might leverage from its community:

And, so, we went ahead and put the pieces in place, but the core belief was that there were—we only have about thirty principal developers at this point—and the belief was that there would be hundreds if not thousands of people who would be willing to contribute development time and that it would therefore be unprincipled of us not to allow those people to do so. So, that's what we're trying to do. And, the early results, as you probably know, are quite positive. We're getting patched submissions now every day that we're folding into the code....I don't think we've had enough time beyond the release yet of the source code to characterize how many additional virtual developers we have, but it's already more than a few.

So in the end, Second Life will have given away the copyright to the creativity built in the game, and it will have opened the client and servers to the free sharing of a GPL'd software project. In short, it will have taken just about all its assets and freed them to the community. And from these gifts, it expects to inspire a creativity that will make the platform extraordinarily valuable.

It's been about seventeen years since the World Wide Web became more than a dream of Tim Berners-Lee. As it has seeped into our veins, it has changed how we interact. More of us do things for other people, even if we do it because it is fun. More businesses find ways to do things for us, because doing so is more profitable for them. And more are experimenting with ways to build value by working with a community—commercial economies leveraging sharing economies to produce hybrids.

What has driven companies to the hybrid experiment? The answer is the same as it always has been, within competitive markets at least: the recognition of success. Companies see the hybrid as a model of success, not as a compromise of profit. Tapscott and Williams summarize the lesson for companies building a hybrid consistent with the principles of "wikinomics" (openness, peering, sharing, and acting globally):[59]

> What was the difference [between successful business and not]? The losers launched Web sites. The winners launched vibrant communities. The losers built walled gardens. The winners built public squares. The losers innovated internally. The winners innovated with their users. The losers jealously guarded their data and software interface. The winners shared them with everyone.[60]

Hybrids are also a model for getting others to innovate in ways that benefit the company. Tapscott and Williams quote Technorati's Tantek Çelik: "It comes down to a question of limited time and, frankly, limited creativity. No matter how smart you are, and no matter how hard you work, three or four people in a start-up—or even small companies with thirty people—can only come up with so many great ideas.[61]

Recognition alone is not enough. Remaining is the hard problem of building the community to support the hybrid. So how does community get built?

The answer is: explicitly and upfront. "You have to design with the community in mind," Jerry Yang said. "It's very hard to do . . . as an afterthought. . . . Community has to be part of the product."

But communities can't simply be called into life. Instead, they have emerged through a number of stages. "The first couple elements," Yang told me, were when "the users created content and the users distributed content." But that alone could not sustain a community. "Really critical for a sustainable kind of community," Yang said, "is the community themselves as editors—both as to quality and for the community to continue to have its identity." These together meant the community was more than a bus stop (a place convenient for people to hang out on their way to doing what they wanted to do). Instead, it was a place where the members felt a kind of ownership. And that pride of ownership meant they took steps to improve what they found.

The real challenge, however, comes now, as these online communities begin to offer more. As Yang put it, "People are starting to really put the economic element in. . . . [They are] starting to make money from this community, whether it is printing calendars or doing albums or whatever."

How this gets done well is not something even Yang fully sees yet. "There's no science in this . . . there's still a lot of art."

You know, the community will give you a certain latitude for facilitating a commercial business, but they ultimately have to feel like they're the ones that are part of the economy and not you. So you have to find a way to balance out the value that you as a platform for that community are adding versus the ability for the community to benefit from that economy.

"The goal is to make people feel like there are real social benefits and economic benefits for them to invest in this platform." And in an analogy that we will return to at the end of the book, Yang continued: "It's not unlike running a government where you're charging for certain basic services. If you overcharge, people feel like they're overtaxed, and if you don't charge enough then you don't provide enough services. So it's always a balancing act."

To say that community is important—or in the terms I've been suggesting, to say the "sharing economy" is important—is obviously not to say community alone is enough. Nor is it even to say that among all the factors of success, community is the most important thing. As perhaps the Net's most famous publisher, Tim O'Reilly, put it,

> You get your initial enthusiasm from people who are your passionate, inner core. And you want to build a system that lets them become passionate, become connected, have this personal connection to you and what you do. But, to get beyond that, you have to build a system in which people participate without knowing that they participate. That's where the power comes.[62]

For, O'Reilly put it, the most powerful architecture and participation don't rely on volunteerism.

> They don't rely on people explicitly being part of a community or knowing who's in the community....[T]he Web is like that.

It's this brilliant architecture where everybody puts up their own Web site for their own reasons, links to other people for their own reasons, and yet, there is a creation of shared value.

O'Reilly points to YouTube as an example. YouTube's success, he argues (agreeing again with Chen), came not from the rah-rah of the community activism. Its success came instead from great code: YouTube's success, O'Reilly explained to me,

> wasn't because [people] thought it was cool. It was because YouTube figured out better how to make it viral. Viral is about making it serve the people's own interest, so that they're participating without thinking that they're participating. Google said, "Upload your video here and we'll host it," and YouTube with their Flash player said, "Either put this video on your site or we'll host it anyway." So you get to share the video without any of the costs and with no-muss-no-fuss.

"People aren't thinking," he continued, "that they are donating to YouTube. They're actually thinking, 'Wow! I'm getting a free service from YouTube.'"

O'Reilly's point is a good one. It builds directly on Bricklin's. You create value by giving people what they want; you create good by designing what you're offering so that people getting what they want also give something back to the community. No one builds hybrids on community sacrifice. Their value comes from giving members of the community what they want in a way that also gives the community something it needs. The old part in this story is that in a competitive market, success comes from satisfying customers' demands. The new part is to recognize a wider range of wants, some me-motivated, some thee-motivated, and the way technology can help serve them.

EIGHT

ECONOMY LESSONS

When commercial and sharing economies interact, they produce the hybrid. How they sustain a hybrid successfully is a harder question. We've not yet seen enough to say anything conclusive. We have seen enough to describe a few important lessons.

Parallel Economies Are Possible

The simplest but perhaps most important conclusion is that parallel economies are possible. Work successfully licensed in a commercial economy can also be freely available in a sharing economy. If this weren't true, then there would be no commercial record industry *at all:* despite the war on file sharing, practically every bit of commercially available music is also available illegally on p2p networks; this "sharing" has not been stanched by either the war against it waged by the recording industry, or the Supreme Court's declaring the practice illegal.[1]

Yet despite this massive sharing, according to the recording industry's own statistics, sales of music have declined by 21 percent.[2]

If parallel economies were not possible, that 21 percent would be 100 percent.

Voluntary parallel economies are also possible and often profitable. When labels discovered artists in ccMixter and then signed them to record deals or contracts, the work the artist had freely licensed continued to be free. Indeed, sometimes the very same song was licensed both commercially and noncommercially. This helped the commercial. More artists and record companies will do the same in the future.

Tools Help Signal Which Economy a Creator Creates For

As creators choose between these two economies, the market has developed—free of government intervention—tools to signal which economy they intend to be a part of. When they want to be part of the commercial economy exclusively, they have associated with the traditional representatives. The RIAA, for example, speaks well for those artists who want their art distributed according to the rules of a commercial market only. "All Rights Reserved" is the familiar refrain. "Don't share" is the unfortunately familiar slogan. But when artists want to create for the sharing economy, increasingly they use signs that mark them as members of this economy. Tools such as the Creative Commons "Noncommercial" license enable an artist to say "take and share my work freely. Let it become part of the sharing economy. But if you want to carry this work over to the commercial economy, you must ask me first. Depending upon the offer, I may or may not say yes."[3]

This sort of signal encourages others to participate in the shar-

ing economy, giving them confidence that their gift won't be used
for purposes inconsistent with the gift. It thus encourages this sort
of gift economy—not by belittling or denigrating the commercial economy, but simply by recognizing the obvious: that humans
act for different motives, and the motive to give deserves as much
respect as the motive to get.

Crossovers Are Growing

As the second lesson suggests, nothing bans crossovers. There's
nothing wrong, for example, with an artist who created something
she offered the sharing economy for free then taking that creative
work and selling it to NBC or Warner Bros. Records. Indeed, this
happens all the time in the world of CC-licensed work. Because
economies can be parallel, many participants have discovered that
playing in one economy does not disqualify you in the other.

In 2005, for example, a Los Angeles–based comedy collective
called Lonely Island was trying—like many such collectives—to
get discovered. It posted all its material to the Web with a Creative
Commons license, enabling others to share its work and remix its
work, so long as they gave credit back to Lonely Island. For example,
the collective shot a pilot for Fox called *Awesometown*. Fox rejected
it, but Lonely Island posted the pilot in full on the Web under a CC
license. The collective used the license both to encourage the spread
of its work and, as its members commented in an interview, to "protect ourselves and our fans. That's what sold us on it. It lets everyone
know that they are free to share and remix our stuff, all the rules
are right there—they don't even need to ask permission."[4]

Someone at *Saturday Night Live* saw the group's work and loved it. In the fall of 2005, one member of the collective joined *SNL* as a cast member; the other two joined as writers. Their work continues to be available under the CC license. But the licenses also helped them cross over to a commercial economy.

Strong Incentives Will Increasingly Drive Commercial Entities to Hybrids

Their rhetoric notwithstanding, hybrids are in it for the money. Commercial entities leveraging sharing economies do so because they believe their product or service will be more valuable if leveraged. And sharing economies that bring commerce into the mix do so because they believe revenues will increase. The hybrid is a way to produce value. If it doesn't, it shouldn't be a hybrid.

The hybrid produces value in part by freely revealing information. By its nature, a hybrid can't control exclusively the knowledge or practice of the sharing economy it builds upon. The design thus leaves the door open on its research or development. The company simply gives this asset away.

Some are puzzled by the idea that giving something away might be a strategy for making more money. Indeed, our intellectual biases about concepts like property lead us almost naturally to believe that the best strategy to produce wealth is to maximize control over the assets we have, including (and most important here) intellectual-property assets.

If you find yourself attracted to this view, then you should survey an increasingly significant field of writing about the opposite strategy, used voluntarily by those seeking the same end: wealth. As Eric von Hippel describes, "Innovations developed at private cost are often revealed freely, and this behavior makes economic sense for participants under commonly encountered conditions."[5] For example, "after the expiration of the Watt patent, an engineer named Richard Trevithick developed a new type of high-pressure engine in 1812. Instead of patenting his invention, he made his design available to all for use without charge."[6] The work spread to be foundational in the field. Von Hippel's work provides a host of modern examples that follow the same strategy.

These innovators reveal not as an act of charity, but as a strategy to better returns. "Active efforts to diffuse information about their innovations suggest that there are positive, private rewards to be obtained from free revealing."[7] Put differently, intentional "spillovers" of information may often benefit both the public and the private entity making the spillover.

Again, to many, this may feel counterintuitive. Economics teaches us that a spillover (meaning a resource that is made available to entities that didn't contribute to its production) is an externality (albeit, a positive externality). Externalities, the lesson goes, should be "internalized." Whether it is positive (think: beautiful music) or negative (think: pollution), the person creating the spillover should pay for it—whether positively (by cleaning up the pollution), or negatively (by collecting some reward for the good produced to the public). As von Hippel writes,

> The "private investment model" of innovation assumes that innovation will be supported by private investment if and as innovators can make attractive profits from doing so. In this model, any free revealing or uncompensated "spillover" or proprietary knowledge developed by private investment will reduce the innovator's profits. It is therefore assumed that innovators will strive to avoid

spillovers of innovation-related information. From the perspective of this model, then, free revealing is a major surprise: it seems to make no sense that innovators would intentionally give away information for free that they had invested money to develop.[8]

But in fact, historically, spillovers have produced great value to society. William Baumol estimates that "the spillovers of innovation, both direct and indirect, can be estimated to constitute well over half of current GDP—and it can even be argued that this is a very conservative figure."[9] The Internet is the simplest example: the "inventors" of the Internet captured a tiny fraction of its value; the spillover has been critical to most economic growth in America over the past fifteen years.

That good is enjoyed not just by the society. Instead, it is also often (though certainly not always) enjoyed by the person revealing the information. Describing authors of open-source software, for example, von Hippel writes,

> [If] they freely reveal, others can debug and improve upon the modules they have contributed, to everyone's benefit. They are also motivated to have their improvement incorporated into the standard version of the open source software that is generally distributed by the volunteer open source user organization, because it will then be updated and maintained without further effort on the innovator's part.[10]

These considerations lead many to conclude, with Baumol, that "despite the substantial spillovers (externalities) of innovation, expenditure on R&D in the free-market economies may nevertheless be quite efficient."[11] Put differently, of all the problems we need to solve, eliminating positive externalities should perhaps be quite low on our list.

So when does it make sense to reveal information in order to build a hybrid? One factor is the heterogeneity of the customers. "Data are still scanty," von Hippel writes, "but high heterogeneity [or diversity] of need is a very straightforward explanation for why there is so much customization by users: many users have 'custom' needs for products and services."[12] Another factor is the feedback companies get from this kind of collaborative innovation. Again, von Hippel: "the commercial attractiveness of innovations developed by users increased along with the strength of those users' lead user characteristics."[13] Encouraging these "lead users" to innovate is thus a powerful way to push innovation at the firm.[14]

While hybrids will increase with the spread of the Net, I am not describing some special rule of economics that lives just in the virtual world. Indeed, to the extent that the hybrid is spreading the right to innovate, the dynamic is again following the very old principle I described above: shifting innovation out of the core of the corporation where transaction costs permit.

The hybrid teaches us that this strategy will increase as technologies for reducing transaction costs proliferate. And conversely, it would be checked by changes that increase the transaction costs of the hybrid.

Perceptions of Fairness Will in Part Mediate the Hybrid Relationship Between Sharing and Commercial Economies

We are not far into the history of these hybrid economies. And early enthusiasm will no doubt soon give way to a more measured, perhaps skeptical view. Those contributing to the sharing economy within a hybrid will increasingly wonder about this world where their free work gets exploited by someone else. Should they be paid? How long will these hybrids last?

Some fear it won't last long. At a conference in San Francisco, Rich Green, an executive vice president for software at Sun, expressed doubt. "It really is a worrisome social artifact," Green said. "I think in the long term that this... [is] not sustainable. We are looking very closely at compensating people for the work that they do."[15]

But as we've seen, simply compensating people is not necessarily a solution. The ethic of a sharing economy and that of a commercial economy are different. Were the work of these volunteers plainly part of the commercial economy, then the answer would be easy: of course, they should be paid; and unless they are paid, they will stop the work. And were the work of the volunteers plainly part of the sharing economy, then the answer would be easy as well: you no more pay volunteers than you (should) pay for sex.

If the work of these volunteers is part of a hybrid, however, we don't yet have a clear answer to this question. If the hybrid feels too commercial, that saps the eagerness of the volunteers to work. Brewster Kahle founded the nonprofit Internet Archive after profiting from many commercial enterprises, as he told me: "If you feel like you are working for the man and not getting paid, visceral reactions will come up.... People have no problem being in the gift economy. But when it blurs into the for-pay commodity economy... people have a 'jerk reaction.'" A "jerk reaction": the feeling that they, the volunteers, are jerks for giving something to "the man" for free. No sense could be more poisonous to the hybrid economy, yet like it or not, the skepticism is growing. Said Om Malik, founder of GigaOmniMedia:

I wondered out loud if this culture of participation was seemingly help[ing] build business on our collective backs. So if we tag, bookmark, or share, and help del.icio.us or Technorati or Yahoo become better commercial entities, aren't we seemingly commoditizing our most valuable asset—time? We become the outsourced workforce, the collective, though it is still unclear what is the payoff. While we may (or may not) gain something from the collective efforts, the odds are whatever "the collective efforts" are, they are going to boost the economic value of those entities. Will we share in their upside? Not likely![16]

Anil Dash, a vice president with Six Apart, posed this question more directly: "Should Flickr compensate the creators of the most popular pictures on its site?"[17] Tapscott and Williams describe the reply by Caterina Fake, cofounder of Flickr:

[T]here are systems of value other than, or in addition to, money, that are very important to people: connecting with other people, creating an online identity, expressing oneself—and, not least, garnering other people's attention. She continued on, saying that the Web—indeed the world—would be a much poorer place without the collective generosity of its contributors. The culture of generosity is the very backbone of the Internet.[18]

It is hard for many to see how a "culture of generosity" can coexist with an ethic of profit. The slogan "You be generous, and I make money" seems like a nonstarter. And so increasingly, we must ask how these different norms might be made to coexist. Jeff Jarvis, journalist and blogger, suggests companies "pay dividends back to [the] crowd" and avoid trying too hard "to control [the gathered] wisdom, and limit its use and the sharing of it."[19] Tapscott and Williams make the same recommendation: "platforms for participation will only remain viable for as long as all the stakeholders are adequately and appropriately compensated for their contributions—don't expect a free ride forever."[20]

The key word here is "appropriately." Obviously, there must be adequate compensation. But the kind of compensation is the puzzle. Once again, the "sharing economy" of two lovers is one in which both need to be concerned that the other is "adequately and appropriately compensated for [his or her] contribution." But writing a fat check as "thanks for last night" is not likely to work.

Similarly, there's another important ambiguity in this notion of a "free ride." For again, if Bricklin is right—if the entity succeeds when it is architected to give the user what he wants while contributing something back to either the commercial entity, the sharing economy, or the hybrid—then is the entity really getting a "free ride"? Consider two examples:

- You submit a search query to Google and then click on one of the links Google returns. You have given Google something of value—the information that you judged one link to be the appropriate answer to the query you selected. Google has built its company upon such gifts of value. Is Google riding on your work for free? Or are you riding on Google for free?

- You post a video to YouTube and then embed it in your blog. You have given YouTube something of value—another string of customers further strengthening its market share, drawing viewers in through a channel that helps YouTube understand who they are, and hence advertise to them more profitably. Is YouTube riding on your work for free? Or are you riding on YouTube?

The point should be obvious: both the user and the company benefit from the interaction. And when both benefit, how do we say who is riding for free?

These questions, however, like any questions of perceived justice, won't be decided on the basis of logic alone. They will turn instead upon practices and understandings. There are lines that companies can't cross. Those lines are drawn by the understanding of those within a community. To their communities, hybrids will try to signal their virtue, or the fairness of the exchange they offer. Craig Newmark, for example, emphasizes moderation as a key to his hybrid's success. ("People see that we're not out to make lots of money, and people can see that we've given away a lot of power over the site through the flagging mechanism and how we actually try to listen to people in terms of suggestions and then follow through.") Tim O'Reilly makes a similar point, though not about moderate profits. He emphasizes moderate efforts at control:

> There is a social compact.... [And] people in some sense will regard certain people as good guys and so they'll go further for them than they would for somebody who they regarded as a hostile. A really good example of that is... Safari, our online book service. We have a very light DRM [for those books, making] it hard to spider. But not too hard.... And we get thousands of e-mails from people reporting that to us. You know, "Hey, I found

your books on a site in Russia." "I found your books on a site in Romania." I bet the RIAA didn't get those e-mails.

Philip Rosedale of Second Life emphasizes a different value: transparency. "I'm a tremendous believer in the idea that there's a new mode by which businesses can interact which is based on complete transparency," he explained to me.

> There's a trust-building exercise there that doesn't traditionally happen, because companies are inherently private because historically, competition was the first order of concern of companies. Therefore privacy [or secrecy], for the purpose of giving you a leg up on your competitors, has always been a kind of a central building block of corporate behavior. And, in a world where openness and network effects are likely to decide the winner, you now have to break down that perception. You want to build a company whose first value is not privacy but, instead, disclosure.

Some companies go even further. Brewster Kahle describes the decision of the search company Alexa.

> [W]e wanted to build a new-generation search engine, which is sort of what Alexa and the Internet Archive strove to do in '96. It turns out that we were wrong, that the world didn't need a fundamentally different kind of search engine, because the search engines were going along okay. But Alexa would collect the World Wide Web and make a service based on it. So that it was a for-profit company that leveraged the community work of others, which was the contents of the Web, to produce a service.

But exploiting that material beyond being directly tied to producing that service, we at Alexa did not feel it was right. So Alexa donated a copy of what it had collected to a nonprofit and deleted its own. So after six months, that was how long it took for Alexa to use the material to do the service that people were, in some sense, cooperating [with] or allowing Alexa to have built. It donated a copy, and it deleted theirs. We think of that as very … we thought of that as a very important part of the balance of the property interest in a commercial company and the public interest that can be better served within a nonprofit.

Every company building a hybrid will face exactly the same challenge: how to frame its work, and the profit it expects, in a way that doesn't frighten away the community. "Mutual free riding" will be the mantra, at least if the value to both sides can be made more clear.

There are of course famous examples of this mutual free riding gone bad. The volunteer-created CD Database—CDDB— was one.

As I've already described, CDDB was an online database that contained track information about CDs. That information was not included on the CD itself. Instead, it was added by users. The inventor of CDDB, Ti Kan, thought "[I]f I typed in all this information about a given CD, why should Joe down the block have to type in the same information," as David Marglin, general counsel of the company that eventually took over CDDB, Gracenote, explained to me. So Kan and his collaborator, Steve Scherf, "started figuring out a way to collect various other people's collections as typed in by them. And then that repository became the CDDB."

All this typing was done voluntarily. People wanted their machines to know what the tracks were; they were happy to share with others the information they typed into their machines. And Kan and Scherf built the tools to aggregate the results of this voluntary work "with the best intentions in the world. They were not trying to make any money off of it. They really wanted to make just a social network and get all the network effects." They built a commons for others to add to; volunteers demonstrated the "grace of the commons" through the contributions they made.

But, Marglin explained to me, as more and more people began to rely upon this database to identify their CDs, Kan and Scherf began to realize "very quickly that they had a beast on their hands because: one, in order to be any good, the software would have to do a lot more reconciling than they first expected...and two, the amount of server space that they would need in order to take in, not only all the lookups, but...all the submissions, was just going to overwhelm them."[21]

So the founders of CDDB started looking for a way to make sure their creation would survive.

Gracenote would become their savior. With the help of a serial entrepreneur, Scott Jones, Tan and Scherf launched the for-profit Gracenote, to generate the revenues necessary to support the database, and to profit from the idea they had brought to life. Gracenote started licensing the database to whoever would pay for it. And soon many started to wonder why this thing of value, created through the work of volunteers, could be so easily sold. Many complained that Gracenote took "what [was] essentially an open-source database and clos[ed] it off."[22] There was an explosion of criticism within the online community.

Gracenote wasn't prepared for the criticism. Marglin told me that there was a lot of "bad mojo in the air from people who didn't understand the transition from 'here's a couple guys in the garage' to this 'it's a world-class service that has to be able to power Apple software.'" More transparency would have helped here, as would a clearer way for the collaborators to benefit. Again, Marglin: "The compensation may not be dollars but maybe credits or attribution for something that is of value so that there is a value exchange between the conduits."

It was the shift in rules, against the background of different expectations, that produced a fairly strong reaction against Gracenote—even though to this day, Gracenote still has "a program where if you're a programmer and want to develop a noncommercial application and don't want to derive any revenue from it, you're free to get and use Gracenote software."

Many believe this commercialization of a free project weakens the incentive for volunteers to contribute to it. Brewster Kahle, for example, holds this view. Gracenote became less after its founders made more. The same is true for a parallel database, also initially built by users for free, the Internet Movie Database, or IMDb. As Brewster explained to me,

> Both [CDDB and IMDb] went off into commercial organizations. And there was a feeling of betrayal on the part of those who contributed their efforts for free. Because somebody else seemed to be taking advantage of them or not offering it back to the commons.
>
> Now with IMDb, Amazon is thrilled that they are paying their own bills based on their advertising and subscription-based

premium products.... But did IMDb end up what it could have
been? I don't think so.

The conflict here is deeper than the Internet. It will get resolved
in the hybrid economy only when each economy—the commercial and the sharing—validates the other. If those within a sharing
economy hate commerce—if they're disgusted with the idea of anyone profiting, anywhere—then the prospects for healthy hybrids
are not good. Likewise, if those within the commercial economy
ignore the ethic of sharing—if they treat members of the sharing
economy like customers, or kids—then the prospects for a healthy
hybrid are not good either.

Not surprisingly, copyright is an important tool in mediating
these expectations. As a fantastic report by the analysts Pike &
Fischer argues,

> Strict copyright enforcement on user-submitted/generated content
> may also result in a negative shift in attitude in the communities
> that build around all these sites. The widespread appeal of user-
> submitted/generated content, especially within the framework
> of social networking sites, relies strongly upon the freedom of
> expression, which often circumvents copyright law. Any further
> restriction to that expression may result in a reduction of user activ-
> ity—the lifeblood of an advertising revenue–based business model.
> In the end, copyright holders and social networking sites may be
> forced to strike a balance in order to maintain user-interest.[23]

The norm that would encourage a hybrid economy most is
the norm of many within the GNU/Linux community: any use,
including commercial use, is fair. The problem with this norm

in the context of culture (as opposed to software) is that cultural products are less obviously collaborative. The feeling of individual exploitation is therefore much more likely to be real. When Google takes the free labor of the GNU/Linux community and leverages it into the most successful Internet business in the world, no member of that community could reasonably say, "Google has exploited me!" But if Sony took a song licensed in a similarly free way and sold 1 million CDs without giving the creator a dime (again, plainly possible under a copyleft license), there's little doubt that someone would have a strong claim of exploitation.

More problematically, free software rests ultimately upon a shaky economic foundation. Robert Young, for example, following the GNU Manifesto,[24] explains the free-software ethic as a version of the Golden Rule: "There's no ideology. There's no complexity. It's do unto others as you would have them do unto you. And if they're doing unto you what you are doing for them, they're welcome to use your technology for whatever they want to use it for because you're receiving a benefit from them." This states a beautiful ethical principle. But it doesn't quite constitute an economic motivation. No doubt "you're receiving a benefit from them." But you could receive that benefit whether or not you contributed your technology back. It might well be the "fair thing" to give back because you've taken. I certainly hope my kids think this way. The hard part is believing that over time, many will continue to believe that giving back makes sense.

Or more skeptically still: There are lots of reasons to believe that the particular character of free software makes it rational to keep the code free—for example, the costs of synchronizing a private version often overwhelm any benefit from keeping the code private.[25] IBM, for example, was free to take the Apache server and build a private label version that would sell, without releasing to others any improvement it made in the code. But that benefit would have been purchased only by IBM's continuing to update its code to reflect changes made in the public version of Apache. At first (when the code bases are close), such updating is not too hard. But over time (as the code bases diverge), it becomes increasingly costly to maintain the private code. Thus the purely rational strategy for this kind of creativity is to innovate in the commons, since the cost of innovating privately outweighs the benefit.

But that story hangs upon the physical characteristics of complex coding projects. Those same characteristics don't exist with, say, a song or a novel. Whatever incentive there may be to stay in the commons with complex, collaborative goods, that same incentive doesn't carry over automatically to all creative works. For these works, nothing more than a norm will support keeping the resource open. And whether, or how, that norm survives is not, to me at least, clear.

It is clear to me, however, that we must avoid a kind of intellectual imperialism. We must be open to the differences in cultural goods—software versus movies; music versus scientific articles. The norms that support free production in one are not necessarily the norms that support free production in the other. As Brian Behlendorf, a cofounder of the Apache project, puts it,

> I don't know if there is one single, social contract amongst cultural works. I think it's different for DJs and electronic music than it is for folk music or even different types of dance music. [For example, the] R & B community has appropriated a lot of tracks and a lot of song examples yet probably would fight fiercely with the RIAA against downloading of their work and such.

He points to another important difference as well:

> A lot of software is, by its nature, a team effort with lots of
> iterations over time. A lot of cultural media—songs or plays or
> movies—are, if they have teams, much smaller teams where they
> tend to be the vision of one or two people. And they arrive at a
> finished body of work or mostly finished work and then it's kind
> of put in a time capsule. . . . [M]aybe it's this evolutionary nature
> of software, the fact that contributors come and go over time and
> that, sure, you have certain versions that are well known and such,
> but it's not like movies where there's a finished work that makes its
> way to the theater and people pay money to go see it.

We need to understand these differences to envision the lines
communities will draw.

"Sharecropping" Is Not Likely to Become a Term of Praise

Although I am uncertain about how these norms will develop
generally, there are particular conclusions we can draw with confidence. Here's one: digital sharecropping will not be long for this
world. Of all the terms that creators from the sharing economy will
demand, the right to own their creativity will be central.

"Sharecropping"?

In April 2007, an excellent research assistant did a survey of
every site she could find that invited creators to "remix" or "mash
up" content provided by the site. There were over twenty-five sites
in her survey. But as she read through the terms of these many

remix contests, a key difference among them became apparent. In 56 percent of the sites, the artist or remixer owned the rights to his remix—not the content he remixed, but the remix. Sixteen percent of these sites required that the remix be licensed under some form of public license (okay, all of those that did require a license required that it be a Creative Commons license). But in 28 percent of these sites, the artist or remixer got no rights in her work at all. She was the creator, but she did not own her creativity. She was, in the totally neutral, noninflammatory terminology that I've selected here, a sharecropper.

David Bowie's contests were typical. Bowie's contract required that the remixer "grant, sell, transfer, assign and convey...all present and future rights, title and interest of every kind and nature" in the remix. The remixer also waived any moral rights he might "feel" he had in the remix. And the remixer granted to Bowie's label, Sony, "worldwide royalty-free, irrevocable, non-exclusive license" to any content added to the Bowie content to make the remix.[26] Thus, if you composed a track and uploaded to the Bowie site to remix it with something Bowie created, then Bowie was free to take that track and sell it, or use it in his own music, without paying you, the artist, anything.

This trend away from artists owning their creations is not new. It has long been a part of commercial creativity. In America, for example, the "work-for-hire" doctrine strips the creator of any rights in a creative work made for a corporation, vesting the copyright instead in that corporation.

This is an awful trend, fundamentally distorting the copyright system by vesting copyrights in entities that effectively live forever. (For that reason, the copyright given to humans is life plus seventy, but to corporations, a fixed ninety-five years. Unfortunately, unlike humans seventy years after their death, after ninety-five years there is still an "artist" eagerly begging Congress for more.)

Yet whether or not the work-for-hire doctrine flourishes in the commercial economy, my sense—and no doubt, my bias—is that sharing-economy creators will increasingly demand at a minimum that their rights remain theirs.

We've already seen a similar frustration brew in the context of "fan fiction," particularly around the *Star Wars* franchise. As with the Harry Potter story, Lucasfilm learned early on that there were millions who wanted to build upon *Star Wars,* and few who thought themselves restricted by the rules of copyright. Like Warner, Lucasfilm recognized that these fans could provide real value to the franchise. So under the banner of encouraging this fan culture, Lucasfilm offered free Web space to anyone wanting to set up a fan home page.

But the fine print in this offer struck many as unfair. The contract read:

> The creation of derivative works based on or derived from the Star Wars Properties, including, but not limited to, products, services, fonts, icons, link buttons, wallpaper, desktop themes, online postcards and greeting cards and unlicensed merchandise (whether sold, bartered or given away) is expressly prohibited. If despite these Terms of Service you do create any derivative works based on or derived from the Star Wars Properties, such derivative works shall be deemed and shall remain the property of Lucasfilm Ltd. in perpetuity.

Translation: "Work hard here, *Star Wars* fans, to make our franchise flourish, but don't expect that anything you make is actually

yours. You, *Star Wars* fans, are our sharecroppers. Be happy with
the attention we give you. But don't get too uppity."

These terms incited something of a riot among the fans. As one
of their leading spokesmen put it, "The real story is a lot uglier, and
has much less to do with the encouragement of creativity than its
discouragement—there's nothing innocent about Lucasfilm's offer
of web space to fans."[27]

"Nothing innocent," again because all ownership went to
Lucasfilm. Fans may well agree not to profit from their work; they
may well think it fair that any commercial opportunity arising
out of *Star Wars* be ultimately within the reach of George Lucas.
(Remember, this was the good sense expressed by Heather Lawver,
leader of the Potter Wars.) But that concession does not mean the
fans believe that their work too should be owned by George Lucas.
And so when Lucasfilm pushed to the contrary, these fans pushed
back.

Lucasfilm, however, does not seem to be deterred. In 2007, the
company launched a mash-up site to encourage creators to mash up
scenes from the *Star Wars* series with their own music or images
uploaded to the Lucasfilm server. Who owned the mash-ups? No
surprise: Lucasfilm. But here's the part that really gets me: As
with David Bowie's site, if you upload, say, music you have written to be included in a mash-up that you have made, you not only
lose the rights to the mash-up, *you also lose exclusive rights to your
music*. Lucasfilm has a perpetual, and free, right to your content,
for both commercial and noncommercial purposes. Once again:
sharecropping.

I'm not saying that this virtual sharecropping should be banned.
Instead, I am asking which types of hybrid are likely to thrive. A
hybrid that respects the rights of the creator—both the original

creator and the remixer—is more likely to survive than one that doesn't. That's not because everyone will care about the rights of ownership offered by a particular site. They won't. But competition is won on the margin. And the ham-handed overreaching of the typical Hollywood lawyer is just the sort of blunder most likely to spoil the success of an otherwise successful hybrid.

Think again about what the Lucasfilm site says to the kids it invites to remix Lucas's work. No one doubts that Lucas's contribution to this site is significant. The *Star Wars* franchise is one of the most valuable in history, extraordinarily compelling and creative, not just to my generation, but to anyone. No one would think that Lucas has any moral obligation to give up ownership to that work (at least during the "limited time" period that its copyright survives). This creativity is rightfully Lucas's.

But when Lucas invites others to remix that creativity, he does it for a reason. A financial reason. He wants to leverage the work of thousands of kids to make his original content more compelling. And, in turn, to make the franchise more valuable.

There can be nothing wrong with that objective, at least from my perspective. This is precisely the manner of every hybrid. And we shouldn't shrink from acknowledging that profit is one objective of those within the hybrid mix.

But though the objective of profit is not a problem, the manner in which that profit is secured can be. The respect, or lack of respect, demonstrated by the terms under which the remix gets made says something to the remixer about how his work is valued. So again, when Lucas claims all right to profit from a remix, or when he claims a perpetual right to profit from stuff mixed with a remix, he expresses a view about his creativity versus theirs: about which is more important, about which deserves respect.

To the lawyers who drafted the agreements I'm criticizing (for I'm sure George Lucas, who is an important contributor to education, had nothing to do with selecting these terms), my concerns will seem bizarre. They've spent their whole career striking deals just like this. The agreements between media companies, or media companies and artists, are not love letters. They do not express mutual respect. A lawyer's job (at least in a commercial economy) is to get everything he can. He is to maximize the value for his client. Social justice is not within his ken.

But these lawyers are the most likely to fail in this new environment. They have developed none of the instincts or sensibilities necessary to build loyalty and respect with those in the sharing economy that their clients depend upon. Like the lawyers working for union busters at the start of the twentieth century, they're proud of precisely the behavior that will cause the most harm to their clients.

Others will teach them—not through lectures or exams, but through the success in the market others will have, and that they (or more accurately, their clients) won't have. Competitive markets will reward right behavior. Too bad for their clients, but lawyers know little of competitive markets. (And tenured law professors know even less.)

The Hybrid Can Help Us Decriminalize Youth

As hybrids for culture have developed, the most successful of these hybrids have learned that encouraging *legal* creativity is the key to encouraging a healthy and successful business. The struggles companies like YouTube have faced to address both legitimate and illegitimate claims by copyright owners have led others to plan for these complexities in advance. blip.tv, for example, explicitly enables users to mark their creative work with the freedoms they intend it to carry. Sony's version of YouTube in Japan, eyeVio, requires uploaders to verify the freedom to share the content they intend to upload.

As this model becomes the norm, it will change the world within which this remix creativity happens. The focus will shift to the creativity alone; the conditions under which that creativity happens—and the "piracy" of its making—will no longer be interesting. The hybrid can pave a way toward legal remix creativity. The incentive of the market can drive a market reform to make this form of expression allowed.

I don't think this market incentive alone will be enough. Policy changes will be necessary as well. But one great thing about democracy in America is that when the market demonstrates the wisdom of a certain freedom, politicians at least sometimes listen. As the market of hybrids becomes even more significant, the freedoms hybrids will depend upon will become more and more salient to policy makers.

PART THREE

ENABLING THE FUTURE

In the twentieth century, RO culture flourished. The twenty-first century could make it better. An extraordinary range of diverse culture could be accessible, cheaply, anytime and anywhere. Access could be the norm, not a privilege.

Before the twentieth century, RW culture flourished. The twenty-first century could make it better too. Digital technologies have democratized the ability to create and re-create the culture around us. Our kids have just begun to show us how they create as they spread culture. We've just begun to see how their understanding grows as the practice of remix spreads, and especially as it becomes part of the hybrid.

Never before has the opportunity for the hybrid been as real. Never has the chance for commercial entities supporting sharing economies, or for sharing economies to exploit commercial opportunities, been as powerful. The opportunity of a hybrid economy is the promise of extraordinary value realized. *How* is not obvious. Certainly not automatically, or easily. But technology has now given us the chance to tap human effort for extraordinary good. Subtle and insightful entrepreneurs could transform that opportunity into real wealth.

Thus the future we could have: a better RO culture, a more vibrant RW culture, and a flourishing world of hybrids. They stand in stark contrast to the dark trade-offs the current war seems to present. They reject the idea that enabling the freedom to create in the RW tradition means giving up on any good from RO culture. They reject the notion that Internet culture must oppose profit, or that profit must destroy Internet culture. They are futures of optimism, and an optimism we should all embrace.

But optimism is not a promise. Or a plan. It is at most a reason to plan. It is a motivation for real promises. And in the face of this promise, we need a plan for making these futures real. Real change will be necessary if this future is to come about—changes in law, and changes in us.

NINE

REFORMING LAW

Copyright law regulates culture in America. Copyright law must be changed. *Changed,* not abolished. I reject the calls of many (of my friends) to effectively end copyright. Neither RW nor RO culture can truly flourish without copyright. But the form and reach of copyright law today are radically out of date. It is time Congress launched a serious investigation into how this massive, and massively inefficient, system of regulation might be brought into the twenty-first century.

Providing that comprehensive plan is not my purpose in this book. Instead, in this chapter, I sketch five shifts in the law that would radically improve its relation to RW creativity and, in turn, significantly improve the market for hybrids. None of these changes would threaten one dime of the existing market for creative work so vigorously defended today by the content industry. Together, they would go a long way toward making the system make more sense of the creative potential of digital technologies.

1. Deregulating Amateur Creativity

The first change is the most obvious: we need to restore a copyright law that leaves "amateur creativity" free from regulation. Or put differently, we need to revive the kind of outrage that Sousa felt at the very idea that the law would regulate the equivalent of the "young people together singing the songs of the day or the old songs." This was our history. This history encouraged a wide range of RW creativity. And even if the twentieth century lulled us into a couch-potato stupor, there's no justification for permitting that stupor to sanction the radical change that the law, in the context of digital technologies, has now effected—the regulation, again, of amateur culture.

That regulation could be avoided most simply by exempting "noncommercial" uses from the scope of the rights granted by copyright. No doubt that line is hard to draw. But the law has already drawn it in many different copyright contexts. Eight sections of the Copyright Act explicitly distinguish their applications based upon the difference between commercial and noncommercial use.[1] A jurisprudence could develop to help guide the distinction here as well.

This exemption should at least be made for noncommercial, or amateur, remix. Consider, for example, the following table:

	"COPIES"	REMIX
PROFESSIONAL	©	©/free
AMATEUR	©/free	free

The rows distinguish between professional creativity and amateur. The YouTube video of Stephanie Lenz's eighteen-month-old is amateur creativity; DJ Danger Mouse's remix of the Beatles' *White* and Jay-Z's *Black* albums is professional creativity. The columns distinguish between remix and non-remix, or what I call "copies." "Remix" here means transformative work. "Copies" mean efforts not to change the original work but simply to make it more accessible.

With this matrix then, we can now see at least one clear example of where culture should be deregulated—amateur remix. There is no good reason for copyright law to regulate this creativity. There is plenty of reason—both costs and creative—for it to leave that bit free. At a minimum, Congress should exempt this class of creative work from the requirements of clearing rights to create.

By contrast, copies of professional work should continue to be regulated in the traditional manner. The right to distribute these could, in this model, remain within the exclusive control of the copyright holder.

Professional remix, and amateur distribution, are more complicated cases. There should be a broad swath of freedom for professionals to remix existing copyrighted work; there's little reason to worry much about amateur or noncommercial distribution of creative work—at least if the compensation plan described below is adopted. These categories could thus also be deregulated partially. But neither should be deregulated to the extent that amateur remix should.

What about "fair use"? By "deregulating," I don't mean the doctrine of fair use. I mean free use. Fair use is a critically important safety valve within copyright law. But it remains, perhaps necessarily, an extraordinarily complicated balancing act, and a totally inappropriate burden for most amateur creators. My recommendation is that Congress exempt an area of creative work from the requirements of fair use or the restriction of copyright. It is not that courts find ways to balance the system to free use. By contrast, fair use would remain a critical part of any professional creativity.

But what happens when a commercial entity wants to use this amateur creativity? What happens when YouTube begins to serve it? Or NBC wants to broadcast it?

In these cases, the noncommercial line has been crossed, and the artists plainly ought to be paid—at least where payment is feasible. If a parent has remixed photos of his kid with a song by Gilberto Gil (as I have, many times), then when YouTube makes the amateur remix publicly available, some compensation to Gil is appropriate; just as, for example, when a community playhouse lets neighbors put on a performance consisting of a series of songs sung by neighbors, the public performance of those songs triggers a copyright obligation (usually covered by a blanket license issued to the community playhouse). There are plenty of models within copyright law for assuring that payment. Collecting societies have long provided private solutions to these complex rights problems. Compulsory licensing regimes—where the law either specifies a price, or specifies a process for determining a price that will govern a particular use of a copyrighted work—have done the same.[2] The aim in both cases is to find a simple and cheap way to secure payment for commercial use. The aim as well, I've argued, should be to avoid blocking noncommercial use in the process of protecting commercial use.

This is not the balance the law currently strikes. Perversely, the law today says the amateur's work is illegal, but it grants YouTube an immunity for indirectly profiting from work an artist has remixed. That is just backwards, and legal reform to reverse it is appropriate.

Most of those who would resist this kind of proposal wouldn't resist it for the money. Hollywood doesn't expect to get rich on your kid's remix. Nor does it have a business model for licensing cheap reuse by cash-strapped kids. But it is worried about reputation. What if a clip gets misused? What if Nazis spin it on their Web site? Won't people wonder why Kate Winslet has endorsed the NRA? (Don't worry. She hasn't.)

This problem comes not, paradoxically, from a lack of control. It comes from too much control. Because the law allows the copyright owner to veto use, the copyright owner must worry about misuse. The solution to that worry is less power. If the owner can't control the use, then the misuse is not the owner's responsibility.

Consider a parallel that makes the point more clearly. As every American should know, for almost a century after the Civil War, segregation continued to stain our ideal of equality. The Supreme Court took an important step toward reversing that fact in 1954, when it ruled state-sponsored segregation unconstitutional. But it wasn't until Congress began to enact meaningful civil rights legislation in the 1960s that equality made any real progress at all.

The heart of that new legislation was the Civil Rights Act of 1964, which, among other things, forbade discrimination in public accommodations, including restaurants, bars, and hotels. Among the many witnesses called to the congressional hearings in support of this federal regulation of "public" accommodations were owners of restaurants and hotels in the South.[3]

This fact at first seems puzzling. Why would the target of an extensive set of federal mandates be arguing in favor of those mandates? Ideology isn't an explanation: these witnesses were not integrationists for reasons of principle or justice. Instead, they wanted the government to force them to do something for economic reasons.

As the witnesses explained, however, in principle, they wanted to serve African Americans. Segregation artificially constrained their market. But if they simply opened their doors voluntarily to African Americans, whites would boycott the restaurants. Inviting blacks in would be seen as pro-black. Pro-black in the South was punished. But if the government removed any choice from the restaurants, then opening their doors to African Americans would be ambiguous. It might be because they were pro-black; it might be because they were simply following the law. According to these witnesses, that ambiguity would stanch any retaliation. Segregation could be ended without them having to pay the price for ending it.

For reasons analogous to the civil rights debates, copyright law gives copyright owners more power than they should want. Put differently, like the Southern restaurant owners, they should want the law to remove some of the rights that the law currently gives them.

Why? Well, think again about Warner and its relationship to its fan culture. Warner has come to see what many now understand: that the obsessive attention of fans makes their franchise more valuable. The "piracy" that happens when a fan takes a clip from a Harry Potter film to post on her Web site is productive. It makes Harry Potter more valuable. Warner reaps the benefit of that increased value.

But as Marc Brandon rightly observed, trademark law puts some obligations on the trademark owner. Which means that technically, Warner needs to worry about which content it authorizes and which content it doesn't. Some of those worries are for good purpose: sites that commercialize Warner content cannibalize Warner's profit. But some of those worries are simply political, like content on a site promoting sex education, or opposing abortion.

Warner must police such uses because not doing so might be seen as an endorsement of the particular uses. The fact that it is

Warner's right means that it is also Warner's responsibility. Any mother opposing a particular way that Warner's material is used would be right to say to Warner, "This is your fault because you're perfectly able to say "no" to this."

Just as the Southern restaurant owner was free to say no to black patrons. That freedom meant responsibility, even when the responsibility had nothing to do with the ultimate purpose or profit of the restaurant.

Were the law to curtail Warner's rights by exempting from copyright regulation any noncommercial use of creative work, however, then Warner would not be responsible. When a parent objected to the use of Harry Potter on a site that also promoted Republican/Democratic ideals, Warner's perfectly fair response would be, "There's nothing we could do about that. We don't have the right to regulate noncommercial use. This site is plainly using the content noncommercially." Warner's responsibility would thus end where its rights ran out. Its obligation to keep its products pure would be limited to those contexts in which commerce affected the use of its products.

That limit would not just remove Warner's responsibility. It would also increase Warner's profits—at least for work that inspires a fan community. For with the removal of a legal barrier to fan action, more fans will participate in that fan action. And the more who devote their efforts toward Warner creative products, the better it is for Warner.

Less control here could mean more profit. Removing the right to some of that control would thus be a first, and valuable, change that Congress could make toward enabling the RW hybrids to flourish.

2. Clear Title

When Google announced its plans to digitize—or Google-ize—18 million books, the editors at the *Wall Street Journal* were outraged. "There's a happy-go-lucky vibe around Google," the *Journal* wrote, contributing to an "image" that lends a "spin of respectability and beneficence to projects such as Google Print." "But," the *Journal* warned, "the mere activity of digitizing and storing millions of books...raises a serious legal question." Google was ignoring these questions, the *Journal* charged. "Intellectual property was important enough to the Founding Fathers for them to mention it explicitly in the Constitution. We assume that when Google says 'Don't Be Evil' this includes 'Thou Shall Not Steal.'"[4] (Actually, the Constitution doesn't explicitly mention "intellectual property." It speaks of "exclusive rights" to "writings and discoveries"—aka, monopolies. To say that means the framers endorsed IP is like saying they endorsed "war" because the Constitution mentions that as well.)

In this case, the editors missed a fundamental fact about the "property right" that copyright is. Copyright is property. But as currently constituted it is the most inefficient property system known to man. That inefficiency is the core justification behind Google's claim to be allowed to use this content freely. "Fair use," in other words, turns upon this "inefficiency."

Consider a few statistics. Of the 18 million books that Google intended to scan, 16 percent are in the public domain and 9 percent are in print, and in copyright—meaning 75 percent are out of print yet presumably within copyright.[5]

What does it mean to be out of print? The most important practical consequence is that it is virtually impossible to identify who the owners of copyrights are for works that are out of print. Impossible precisely because the government has so totally failed to keep the property rights for these copyrights clear. There's no registry for identifying the owners of copyrighted works. Nor is there even a list of which works are copyrighted. For anyone trying to make culture accessible in exactly the way Google has, the existing system makes it impossible—at least if permission is required for any particular use.

This problem is not limited to Google. Consider the University of Houston's Digital Storytelling project. As I described, the majority of students at the University of Houston are not native English speakers. Many are immigrants. Most were raised in homes where English was not the mother tongue. This mix creates an important language gap: some students speak English significantly better than others, meaning any task exclusively in words is a task that burdens some more than others.

To respond to this difference, Houston began a digital storytelling project. Students were invited to develop short videos that told a story about some period in American history. The stories were to mix images and sounds from the period, in a way that brought the history to life. But as these students discover recordings from the Depression, or photographs of the Korean War, what are they allowed to do with them? I'm not talking about Disney films, or the collected works of America's greatest jazz musicians. I'm talking instead about obscure works whose owner could never be found.

The university's lawyers had a simple answer: they weren't happy with this permissionless use, but they would ignore it so long as the project didn't let anyone see the work. These kids were allowed to create. But what they created could be viewed by no one except their teacher.

Here again, this makes no sense. No one loses because of these kids' use. The law should not inhibit it at all. Yet the uncertainty that now dominates copyright law makes this, and a thousand other projects, uncertain. For no good copyright-related reason.

Digital technologies make it feasible—for the first time in history—to do what Jefferson dreamed of when he founded the Library of Congress: "to sustain and preserve a universal collection of knowledge and creativity for future generations." The costs of digitizing and making accessible every bit of our past are increasingly trivial. At least, the technical costs are trivial. The legal costs, on the other hand, are increasingly prohibitive. Uncertainty destroys the potential of many of these projects. It reserves others to those companies only that can afford to bear this extraordinary uncertainty.

This reality is ridiculous. The main function of copyright law is to protect the commercial life of creativity. Though there are exceptions, in the vast majority of cases, that commercial life is over after a very short time.[6] There is no good copyright reason for the law to interfere with archives or universities that seek to digitize and make available our creative past. And yet the law does.

This problem could be fixed relatively easily by applying an innovation as old as American copyright law.

For most of the history of copyright in America (186 out of almost 220 years, to be precise), copyright was an opt-in system of regulation. You got the benefit of copyright protection only if you asked for it. If you didn't ask in the appropriate way—if you didn't register your work, mark it with a copyright symbol, deposit it with the Library of Congress, and renew the copyright after an initial term—your work passed into the public domain. That fact didn't bother most who published creative work in America. For much of that history, the majority of published work was never copyrighted. The vast majority of work that was copyrighted didn't renew its copyright after an initial term.

This system was cumbersome, and expensive. Sometimes it resulted in an unfair forfeiture of rights. But it had an important benefit—both to other creators, and to the spread of creativity generally. The system was automatically self-correcting. It automatically narrowed its protection to works that—from the author's perspective—needed it. And it left the rest of the world of published creativity free of copyright regulation.

This system of opt-in copyright was abolished beginning in 1976. Motivated in part by the desire to conform to international conventions, and in part by a desire simply to make the system simpler, Congress inverted the old system. Copyright was now an opt-out system, where the regulation of copyright was automatically extended to all creative work upon its creation. No formal acts were required to get the benefit of this protection. And the term was now the maximum term, automatically. When Congress made this change, it probably didn't matter much. Nineteen seventy-six was the climax of the RO culture. Those producing this culture benefited from this automatic expression of the right for them to control.

In the RW era, however, this automatic and fundamentally ambiguous system of property law unnecessarily burdens creativity. And there are obvious, simple ways to change this system to remove this burden. The least destructive change, in my view, would create a maintenance obligation for copyright owners after an initial term of automatic protection. Under one proposal, fourteen years after a work was published, the copyright owner would have to register the work. If she failed to do so, then others could use it either freely or with a minimal royalty payment. The system would clarify rights after fourteen years. The only work that would continue to be fully protected would be work that the author took steps to protect.

This change would thus clear the title to creative work, so the market could regulate access to that work more efficiently. In this sense, the proposal has parallels in every other body of property law. Copyright law is unique in its failure to impose formalities on property owners. In many areas, property law is unforgiving of those who fail to shoulder their fair share of the burden in keeping the system efficient. That principle should be returned to copyright. If it's not worth it for a copyright owner, after fourteen years, to take some minimal step to register her works, then it shouldn't be worth it for the United States government to threaten criminal prosecution protecting the same property.

This change would make the copyright system more efficient. Copyright gives copyright owners a property right—just as real property law gives homeowners a property right to the land on which their house is built. In theory, this is a good thing. By protecting the resource with a property right, the law enables the resource to move to its highest-valued user. The market thus complements this property system by encourage trades to make sure the right is held by the person or entity that values it most.

For that system to function, however, the rights have to be clear. It must be possible to know who owns what. For that reason, for example, owners of land must file a deed describing the metes and bounds of their property. Or the owner of a car must file a registration. These systems are designed to make the market function efficiently. By making claims on the ownership of property clear—or put differently, by making title to property clear ("clear title")—the system assures that the property can be allocated in a way that makes everyone better off. Technology offers an extraordinary opportunity for making registration work efficiently. Already companies such as YouTube are experimenting with technology that can automatically identify video or music in the content that's being uploaded. We could easily imagine a system whereby, as these technologies develop, copyright holders would "register" their work not in the old-fashioned way (by filing a form with the copyright office) but by uploading the works so that servers could take a signature of it, and then add that to the list of creative work monitored for infringement. Works that were added into this kind of "registration system" would get the full protection of the law. Works that were not would, after a period of time, rise into the public domain.

There's lots to question about that theory. But for my purposes here, I want to embrace it without qualification. Assuming one believes the market perfects a well-functioning property system, what has to be true about the property system? Or put differently, assuming we want the market to work well, is the copyright system designed to enable that?

As it is currently built, the answer is no. As currently constituted, copyright is an extraordinarily inefficient property system. Even the *Wall Street Journal* should see that. Or actually, *especially* the *Wall Street Journal* should see that. From its perspective, why would it be surprising that a government-crafted system of monopoly would be inefficiently run?

3. Simplify

The third change follows directly from the second. Congress must
work to make the law simpler.

If copyright were a regulation limited to large film studios and
record companies, then its complexity and inefficiency would be
unfortunate but not terribly significant. So what if Fox has to hire
more lawyers to work through complex copyright licensing problems? (For those of us who make lawyers for a living [law profes-
sors], the need is positively a good thing!)

But when copyright law purports to regulate everyone with a
computer—from kids accessing the Internet to grandmothers who
allow their kids to access the Internet—then there is a special obligation to make sure this regulation is clear. And that obligation is even
stronger when, as here, the regulation is a regulation of speech.

Copyright law fails miserably to live up to this standard. Under
the current law, if a kid wants to legally make a video to post on You-
Tube, synchronizing music from his favorite bands with film clips
from his favorite movies, he has to clear rights. Even if the rights
holders were likely to clear those rights, it would be extremely difficult to track them down. And of course, considering the current
attitudes of the major rights holders, it is impossible to imagine that
they would even entertain the idea of authorizing this remix use.

We thus have a system of technology that invites our kids to be
creative. Yet a system of law prevents them from creating legally.
The regulation of this creativity thus fails every important standard
of efficiency and justice. And Congress should immediately address
how it could be changed to make it work better.

One particular area of the law's failure is the doctrine called "fair use." Fair use is designed to limit the scope of copyright's regulation. A use that would otherwise be within the monopoly right of copyright is permitted by fair use, to advance some important social end. So if my previous books are any indication, there will be many who after reading this book will copy text from it in a highly critical review. Such "copying" technically triggers the law of copyright. But the doctrine of fair use would protect that copying, so long as the scope and purpose of the copying were within the ordinary contours of criticism.

The problem with fair use is not its objective. The problem is how it advances its objective. Once again, the doctrine was developed imagining it would be administered by lawyers. In a world where copyright only (effectively) regulated Fox and Sony Records, that might not be a terrible assumption. But again, when copyright law is meant to regulate Sony and your fifteen-year-old, a system that imagines that a gaggle of lawyers will review every use is criminally inadequate. If the law is going to regulate your kid, it must do so in a way your kid can understand.

Fair use could do its work better if Congress followed in part the practice of European copyright systems. Specifically, Congress could specify certain uses that were beyond the scope of copyright law. Congress should not follow the Europeans completely, however. The flexibility of existing fair-use law does encourage development of the law. It should remain so that those who can afford the lawyers can push the law to develop in ways that make sense of the law. But that system must be married to a clearer and simpler system regulating everyone else.

Conservatives should not resist this point. As one of America's leading libertarian scholars has taught us, the question for regulators is not what rule perfectly advances the policy objectives. The question is whether the return from a complex rule advancing some policy objective is worth the price.[7] In theory, the fancy qualifications and limitations on copyright may well ideally balance protection and incentives. The real question should be whether we couldn't get very close to that ideal balance with a much much simpler (read: cheaper) system.

4. Decriminalizing the Copy

The fourth change is perhaps the most geeky but possibly, in the end, the most important. Copyright law has got to give up its obsession with "the copy." The law should not regulate "copies" or "modern reproductions" on their own. It should instead regulate uses—like public distributions of copies of copyrighted work— that connect directly to the economic incentive copyright law was intended to foster.

To most people, the idea that copyright law should regulate something other than copies seems absurd. How could you have a "copyright" law if it didn't regulate *copies*? But in fact, for most of our history copyright law didn't regulate copies. From 1790 until 1909, the law regulated different uses that directly linked, or were likely to link, to the commercial exploitation of creative work. Thus, it regulated "publishing" and "republishing" of books, as well as "vending" of books—all activities likely to be commercial.[8]

In 1909, the law was changed to refer to "copies."[9] Yet, as I've already described, even that change was not intended to widen the real scope of the law. And in any case, as Lyman Ray Patterson has argued, that change most likely was an error in drafting.[10] Nonetheless, after 1909, the law reached beyond the particular acts that

Congress regulated. The law would reach as far as the technology for "copying" would reach.

The effect of this change in technology was to change radically the scope of copyright law. In 1909, writes Jessica Litman,

> U.S. copyright law was technical, inconsistent, and difficult to understand, but it didn't apply to very many people or very many things. If one were an author or publisher of books, maps, charts, paintings, sculpture, photographs or sheet music, a playwright or producer of plays, or a printer, the copyright law bore on one's business. Booksellers, piano-roll and phonograph record publishers, motion picture producers, musicians, scholars, members of Congress, and ordinary consumers could go about their business without ever encountering a copyright problem.
>
> Ninety years later, the U.S. copyright law is even more technical, inconsistent and difficult to understand; more importantly, it touches everyone and everything. In the intervening years, copyright has reached out to embrace much of the paraphernalia of modern society.... Technology, heedless of law, has developed modes that insert multiple acts of reproduction and transmission— potentially actionable events under the copyright statute—into commonplace daily transactions. Most of us can no longer spend even an hour without colliding with the copyright law.[11]

If copyright regulates copies, and copying is as common as breathing, then a law that triggers federal regulation on copying is a law that regulates too far.

Instead, Congress should adopt again its historical practice of specifying precisely the kinds of uses of creative work that should be regulated by copyright law.[12] The law should be triggered by uses that are presumptively, or likely to be, commercial uses in competition with the copyright owner's use. The law should leave unregulated uses that have nothing to do with the kinds of uses the copyright owner needs to control. Copying, in this world, would not itself invoke federal regulation. Public performances, or public distributions, or commercial distribution, would.

Why is this important? Under the current law, it is easy to get thrown into the briar patch of copyright regulation, and very hard to get out. As I've described, if every single use of culture in the digital context produces a copy, then every single business that purports to use culture in a digital context is potentially subject to copyright's regulation.

In some cases, that's not a problem. It's easy to identify whose permission is required, easy to secure that permission from the copyright owners. But in many other cases, the new or innovative use challenges the copyright holder. The use might create competition, for example, that the copyright holder doesn't want. And so the threat of a copyright-infringement suit against the innovator is an effective way to control that innovation. Once that initial copy is made, the lawyers have to be called into the research lab. Complex and uncertain doctrine is waved around the project. Magic phrases are incanted, as if by witch doctors aiming to ward off inevitable disease.

I know this process well. Though I've never consulted for money in this context, I have been privy to many such conversations among technologists and venture capitalists, trying to understand whether their next great idea will end them up in financial jail (i.e., litigation). I've watched as one side makes a pitch that seems to everyone 100 percent convincing. And then the other side makes a counterpitch that also seems 100 percent convincing. The net result is always a gamble. And because copyright liability is so severe, it is often a "bet-the-company" gamble.

This is an extraordinary waste of economic resources. A business shouldn't need a witch doctor to tell it whether its plan is legal—especially a witch doctor who charges $400 to $800 an hour. The law instead needs to be clearer. And the complication caused by the law being triggered upon the mere creation of a copy is an unnecessary tax on the creative process.

5. Decriminalizing File Sharing

My final change is the one I will describe the least because, again, others have effectively mapped this change.[13] But it is my view that Congress needs to decriminalize file sharing, either by authorizing at least noncommercial file sharing with taxes to cover a reasonable royalty to the artists whose work is shared, or by authorizing a simple blanket licensing procedure, whereby users could, for a low fee, buy the right to freely file-share. The former solution has been described in depth by Neil W. Netanel and William Fisher. The latter is the proposal of the Electronic Frontier Foundation.

While there are advantages and disadvantages to both types of solutions, it is critical that Congress take steps to do one or the other. The reason flows from a simple reality check: a decade of fighting p2p file sharing has neither stopped illegal sharing nor found a way to make sure artists are compensated for unauthorized sharing. In short, the strategy of this decade has failed to advance the objectives of copyright law—providing compensation to creators for their creative work.

No doubt there are still a thousand ideas about how we could regulate the technology of the Internet to kill p2p file sharing. And no doubt there are gaggles of lawyers who would love the easy work of suing kids and their parents for illegal file sharing. But the question Congress should ask is what strategy is most likely to assure compensation to artists, and minimize the criminalization of our youth.

Decriminalizing file sharing is that strategy. As the work of Fisher demonstrates, there are plenty of ways that we might tag and trace particular uses of copyrighted material. That provides the baseline for compensating artists for the use of their creative work. And while I don't believe we need to embrace this system permanently just now, I do believe a transition regulation, designed to compensate and decriminalize, would significantly reduce the collateral damage being caused by the current war.

All five of these changes would go a long way to relieving the copyright system of unnecessary pressure. They would go a very long way to legalizing most of the uses of creative work on the Internet today. And they would not limit in any way the profits of the industries that fight so hard to resist these uses. That's the ultimate test that copyright law should pose: would these uses actually do any harm? The people I spoke with in preparing this book made that point again and again. As Don Joyce explained:

> All lawsuits in this field are always couched in economic terms: "By ripping this off, by stealing this thing, you have threatened the economic viability of the original somehow." They actually say that without any proof that that's true at all. I have never seen anyone prove that in fact, that some work was diminished in its

ability to make money by somebody's sample. In my mind the
work that's reusing it is not in competition with the original. And
you haven't removed the original. It's still there. It will always be
there, as it is, as it was, as the original. And if you sample from it,
you've made something else. You've probably put samples from ten
other things into this thing, and that's just one little element, and
together the parts make up something that has very little bearing
on that one thing you may have sampled from. I just don't think
there's an argument there of what the harm is. If there's no harm,
and it's a lot of fun, I say do it. No one's ever shown me that the
harm is real.

That much is perfectly true. Economists argue ferociously about
whether or to what extent p2p file sharing of complete digital copies of commercially available works might harm the copyright
owner. But no one would even speculate about the harm that comes
from remixing works, much of which is not even commercially
available.

Of course, that's not to say there's no "harm" here at all. As
Johan Söderberg put it to me: "I don't think I'm hurting anybody. I
mean, of course I am hurting [Bush and Blair]. It's something that
is against George Bush and Tony Blair. So I'm hurting them." But
is it copyright's job to protect Blair and Bush from criticism? Is that
the reason the framers granted Congress the right to secure exclusive rights? If you even hesitate to answer that question, then let me
suggest you know very little about the motives of our framers. Giving government the power to silence critics through licensing was
them at their worst (see, e.g., the Alien and Sedition Acts of 1798).
It was not one of the ideals motivating the drafting of our founding
document.

TEN

REFORMING US

The law is just one part of the problem. A bigger part is us. Our norms and expectations around the control of culture have been set by a century that was radically different from the century we're in. We need to reset these norms to this new century. We need to develop a set of norms to guide us as we experience the RW culture and build hybrid economies. We need to develop a set of judgments about how to react appropriately to speech that we happen not to like. We, as a society, need to develop and deploy these norms.

Chilling the Control Freaks

We know the norms this century needs. We can find them if we think again about the freedoms in writing. We were all taught as a kid how to write. We measure education by how well writing is learned. As I've already noted, this is a profoundly democratic feature of our creative culture: we tell everyone they should learn how to speak as well as how to listen.

That core experience brings with it certain expectations—not just about what anyone is free to do, but also about the appropriate response to what anyone writes. Put simply, that response is essentially substantive and laissez-faire. When someone writes a stupid opinion, the appropriate reply questions the opinion, not the right of someone to write it. And while falsity could invite a legal response (defamation), there is (in the American tradition at least) a strong presumption against getting the government involved unless absolutely necessary. The Constitution limits the power of a public figure to sue for defamation. And someone who responds to an error with a lawsuit rightly loses the sympathy of most of us. This is the one place where President Andrew Jackson's mother's advice remains strong: "Never tell a lie, nor take what is not your own, nor sue anybody for slander, assault and battery. Always settle them cases yourself."[1] At least with respect to the slander part, that's got to be right.

As I've already argued, these writing norms are different from the norms that govern much of the arts. The willingness to invoke the legal system to address misuse with films or music is astonishing. As Academy Award–winning director Davis Guggenheim told me eight years ago, it is also increasing.

Ten years ago...if incidental artwork...was recognized by a common person, then you would have to clear its copyright. Today, things are very different. Now if any piece of artwork is recognizable by anybody...then you have to clear the rights of that and pay to use the work. [A]lmost every piece of artwork, any piece of furniture, or sculpture, has to be cleared before you can use it.[2]

We need the norms governing text to govern culture generally. This change must start with the companies that now have legal control over so much of our culture. They must show leadership. Henry Jenkins divides these companies into two sorts:

> [S]tarting with the legal battles over Napster, the media indus-
> tries have increasingly adopted a scorched-earth policy toward
> their consumers, seeking to regulate and criminalize many forms
> of fan participation that once fell below their radar. Let's call
> them the prohibitionists. To date, the prohibitionist stance has
> been dominant within old media companies (film, television, the
> recording industry), though these groups are to varying degrees
> starting to reexamine some of these assumptions.... At the same
> time, on the fringes, new media companies...are experimenting
> with new approaches that see fans as important collaborators in
> the production of content and as grassroots intermediaries helping
> to promote the franchise. We will call them the collaborationists.[3]

These collaborationists are forcing "media companies to rethink old assumptions about what it means to consume media."[4] As they experiment more with freedom, they will encourage norms that support that freedom as well.

Showing Sharing

As I've already described, copyright law is automatic. It reaches out and controls what you create—whether you intend it or not, and whether it benefits you or not. An academic publishing a paper wants nothing more than people to copy and read her paper. But the law says no copying without permission. A teacher with an innovative lesson plan for teaching Civil War history would love nothing more than for others to use his work. The law says others can't without clearing the rights up front. The essence of copyright law is a simple default: No. For many creators, the essence of the creativity is: Of course.

No one needs to question the motive or necessity of those who insist that they must reserve all rights to themselves. Maybe they do. Who am I to say different? But while conceding a necessity sometimes, we should never concede that sometimes means always. That Lucasfilm needs to control all its rights to profit from its genius does not mean that a law professor writing an article about bankruptcy needs the same protection. Or that NBC needs to control the commercial exploitation of *ER* does not mean NBC should have the right to control the exploitation of the presidential debates. The copyright model works well in some places. But some places doesn't mean everywhere.

Movements like the Creative Commons were born to help people see the difference between somewhere and everywhere. Creative Commons gives authors free tools—legal tools (copyright licenses) and technical tools (metadata and simple marking technology)—to mark their creativity with the freedoms they intend it to carry. So if you're a teacher, and you want people to share your work, CC gives you a tool to signal this to others. Or if you're a photographer and don't mind if others collect your work, but don't want *Time* magazine to take your work without your permission, then CC would give you a license to signal this. All the licenses express the relevant freedoms in three separate layers: one, a "commons deed" that expresses the freedoms associated with the content in a human readable form; two, the "legal code," that is the actual copyright licenses; and three, metadata surrounding the content that expresses the freedoms contained within that copyright license in

terms computers can understand. These three layers work together
to make the freedoms associated with the creative work clear. Not
all freedoms, but some. Not "All Rights Reserved" but "Some
Rights Reserved."

In the five years since this project launched, millions of digital
works have been marked to signal this freedom rather than control. Some have used them to help spread their work. Others have
used them simply to say, "This is the picture of creativity I believe
in." And as the tools have been used, they have begun to define
an alternative, privately built copyright system: Almost two-thirds
of the licenses restrict commercial use but permit noncommercial use. The vast majority permit free derivatives, though half of
those require that the derivatives be released freely as well. This is
a picture of a much more balanced regime, built by volunteers, one
license at a time. And it signals something to other artists as well.

This signal is very important, for it shows an alternative that
authors and artists have selected. But more need to show the very
same sign. Whole fields need to establish a different copyright
default. Not necessarily by legislative change. Or at least not yet.
But by the voluntary action of those who believe the default should
be different.

Indeed, if you look at the five changes I suggest copyright law
should make, four of those changes Creative Commons already
enables through the voluntary action of copyright owners.

First, every CC license authorizes at least noncommercial distribution. That goes a great distance in deregulating amateur
creativity.

Second, CC licenses make it simple to identify who a copyright
owner is. More significantly, Creative Commons is now taking the

lead in building an international copyright registry. Both changes help clear title to copyrighted works, and thus help a market in copyright work better.

Third, CC licenses are designed to simplify as much as possible the copyright system it builds upon. Think of the copyright system as the command line interface for computers before Windows or the Mac became common. Like Windows, or graphical user interfaces generally, CC tries to make it easier for the ordinary user to use the copyright system. Not to do everything, but to do the sorts of things ordinary people are likely to need done.

And finally, as every CC license at least authorizes noncommercial copying, we have decriminalized the copy. The question is not "Has a copy been made?" The question is "For what purpose has a copy been made?" This goes a great distance in simplifying copyright in contexts of unexpected or unpredicted uses. That simplification should be Congress's objective as well.

Though I was one of those who helped start Creative Commons, I'm the first to argue that CC is just a step to rational copyright reform, not itself an ultimate solution. But its key advantage is that it works with creators to build a better copyright system. Unlike the standard debate, which sets users against creators, CC is reform advanced by authors and artists themselves. We say what control we need. And in that conversation, we get to debate just how much control is healthy or necessary for a culture.

More creators need to take part in this conversation. More need to ask those who don't why they don't. We all need to work for a norm that doesn't condemn copyright, but rather condemns senselessly deployed copyright. You can be in favor of handguns and oppose giving handguns to kids. Likewise here.

Rediscovering the Limits of Regulation

The final change is perhaps the most important. It is certainly the
most general. We as a culture need to rediscover an idea that was
dominant when Sousa was first learning to conduct: We must recognize the limits in regulation.

We've just left a century in which governmental power across
the world was greater than at any time in human history. So too
were people's expectations for government. At some point in the
course of the century, it became almost natural to imagine that
government could do anything. At some point, it seemed obvious that the only limit to governmental power was governmental
incompetence.

Many in the nineteenth century had a very different view about
government. Many believed that government could do little, or
maybe nothing, to change how people behaved. As the prominent
nineteenth-century legal theorist James C. Carter put it,

> [H]uman transactions, especially private transactions, can be gov-
> erned only by the principles of justice;...these have an absolute
> existence, and cannot be made by human enactment;...they are
> wrapped up with the transactions which they regulate, and are
> discovered by subjecting those transactions to examination. [T]he
> law is consequently a science depending upon the observation of
> facts and not a contrivance to be established by legislation, that
> being a method directly antagonistic to science.[5]

If society is to change or improve, Carter believed, it must do so
by improving the individual. Legislation cannot "originate" that

change, Carter believed, though "it may aid it." "Men cannot be
made better," Carter declared, "by a legal command."[6]

In many ways, my own work could well be characterized as
an apology for regulation. My first book, *Code and Other Laws of
Cyberspace*, argued strongly that while I was as skeptical of our current government as any, it was extraordinarily important that the
government help ensure that the values of cyberspace were our values. My argument was criticized by libertarians, who believed the
best role for government was no role at all. They argued that cyberspace would be best off if government kept its hands off.

I still believe that there are important strategic ways in which
government can do good. But the last few years have convinced me
that we all must be less optimistic about the potential of government to do good.

This (obvious) point first cracked into my head as I was reading the accounts of America's war in Iraq. Whole libraries have
been published about the failures in that war.[7] In book after book,
even those sympathetic to the objectives of the war could barely
find anything good to say about how it was executed. But as I read
more and more of these books, I was struck most by a question that
seemed simply not to have been asked before that war was waged:
What reason was there to think that government power could succeed in occupying and remaking Iraqi society?

I'm not talking about the invasion: that's easy enough. Invasions
are won with powerful tanks and well-placed bombs. I'm talking about everything that would obviously have to be done after
the invasion—from security, to electric power, to food supplies, to
education. It was as if those at the very top simply assumed that the
government could do all these things, without ever asking whether
that assumption made any sense.

What made this all the more weird was that the very people who were operating upon this vision of regulatory omnipotence were the same people who, in a million other contexts, would have been most skeptical about the government's ability to do anything. We're not talking about FDR here. Or even the socialist member of the United States Senate, Bernie Sanders. We're talking about people who don't believe the government can run a railroad. But if a government can't run a railroad, how is it to run a whole society? What possible reason is there to think that we had anything like the capacity necessary to do this?

For though many predicted resistance, the presumption behind our government actions was that force could always quell resistance. If the enemy fought back, we'd fight harder. And at some point, we'd fight hard enough to overcome the enemy. All that was needed was a strong will and good character.

There's a deep fallacy in this way of thinking. In a democracy, more power does not translate into more success. Instead, like a car trying to free itself from a snowbank, in a democracy, more power is often self-defeating. There is a limit to what a government can do that can't simply be overcome by adding power or resources to the problem. At some point, adding more regulation decreases the effective control over the target.

This is not a book about Iraq. But I suggest we can apply the point about that war to the other wars we are waging. There are many such wars that would benefit from such consideration. But the one I want to return to is the war we are waging against our kids because of the way they use digital technology.

Again, as I've described, when p2p file sharing took off, the response of the government (and those who pushed the government) was that this bad behavior should be regulated away. We

assumed that if the government put enough force behind it—
enough prosecutions, enough suspensions from universities—
eventually, the bad behavior would stop.

In fact, the evidence is to the contrary. The government has
passed law after law. It has threatened extraordinary punishments.
And private actors like the RIAA have delivered these punishments
in literally thousands of lawsuits—more than seventeen thousand
as of 2006. But so far, this effort has been a massive failure. Not
because it has failed to protect the profits of the record companies.
No doubt, it has done that to some degree. But the real failure of
this war is the effect that this massive regulation has had on the
basic integrity of our kids. Our kids are "pirates." We tell them this.
They come to believe it. Like any human, they adjust the way they
think in response to this charge. They come to like life as a "pirate."
That way of thinking then bleeds. Like the black marketeers in
Soviet Russia, our kids increasingly adjust their behavior to answer
a simple question: How can I escape the law?

This concern is not just speculation. There is important legal
and sociological evidence to support the concern that overcriminal-
ization in this one (and central) area of our kids' lives could have
negative effects in other areas of their lives, and on attitudes toward
the law generally. To the extent that kids view the laws regulating
culture as senseless, or worse, corrupt, that makes them less likely
to obey those laws. To the extent that they see these senseless laws
as indicative of the legal system generally, they may be less likely to
obey those laws generally. Developing the habit of mind, especially
in youth, of avoiding laws because they are seen to be wrong, or
silly, or simply unjust, develops a practice of thinking that could
bleed beyond the original source. Of course, no one would claim
that laws against piracy increase the incidence of rape or murder.

But there is evidence that if the laws regulating culture are perceived to be morally unjust, that erodes the conditions within a culture for supporting the law more generally.[8]

But I'm not driven to this concern because of compelling T-statistics in a multiple regression. The source of my concern is the literally hundreds of people under thirty who have spoken to me passionately about this issue. Just as I was completing this manuscript, for example, one teenager sent me an essay he had written about "piracy." As his essay, titled "Who Passes Up the Free Lunch," explained:

> You can get any song you want, any type of music, from any era, and it's all FREE! What could be better? All of music's history is at your disposal for one low monthly price: $0!...
>
> I download songs all the time for these exact reasons. It's quick, easy and best of all, free! The other side of it is the artists that lose the revenue that they would have got if all those people had bought the CD's. Obviously not everyone would have bought the CD, but this creates a moral dilemma between supporting the artists and just taking the free lunch. The RIAA says this is illegal, the artists say it's stealing their money, and most of educated society says it's just plain wrong, but here's the problem: I completely agree with both sides.
>
> "Pirating" music, as the RIAA calls it, is something I do whenever I want some new music. All I do is type in the name of the artist or song, and click download, and voila, it appears and starts playing. It's so easy it's like stealing candy from a baby. However, now I'm at a point where I kind of take it for granted, and I don't even think about what I would do if one day it were suddenly not available. In a time when there are so many options to amass a

music collection, I take the way that is most convenient for me and most damaging to the artists. Much has been said in the justice system, in the media and online about the legality of these peer to peer networks, but the reality is that millions of people do it and only a few are ever stopped. I shamefully admit that, despite my sympathy for the artists and others who lose money from the file-sharing, I consciously take part in it and I have no plans to stop in the near future.

Some people can justify stealing music because they do not realize the consequences; that is not the case for me because I am fully aware of them. I live in New York City, a place with liberal political views and that has been the home to many famous musicians over the years. My school, Trinity, is a place full of people with enough money to buy their own CD's, and there is also a club that is trying to encourage people to trade CD's. However, despite many people's support of the philosophy of this group, in this digital age, it is simply infinitely easier to share music online. I even have a guitar teacher who writes songs for a living, and who depends on the money that people steal when they download songs from the internet. I am even so self-consciously guilty about this that I hide the program on my computer so she can't see it when she comes. I tell her that I get my songs from my friends because I know that even if she wouldn't tell me directly, she would be very disappointed in me. Each of these three groups that I am a part of, New Yorkers, Trinitarians, and guitar players, has a moral opposition to "pirating" music, yet somehow, as a member of educated society, I am able to shun this opposition in favor of a "free lunch."

So why is it that I, as well as millions of other people, go on stealing music from artists every day? For some, it may be disrespect for copyright laws, or simply an issue of money that one uses to justify "pirating" music. Others may not know the consequences for the artists, or choose to disregard those consequences. However, there is still another, more hedonistic reason, which I too can identify with, why people still download music from the internet. In a way, it is a product of our take-what-you-can capitalistic society, but at the same time our justice system has said that it is illegal....

The singer "Weird Al" Yankovic also deals with the issue of morality in one of his songs, creatively titled, "Don't Download This Song." He uses his playful lyrics to convince people to buy CD's, but at the same time, he doesn't exactly side with the artists or the RIAA either. He talks about the "guilt" and "shame" that will stay with you after you download the music, but later he talks about the RIAA like it's an evil empire prosecuting children and grandmas willy-nilly. Also, he mentions another part that has been at the back of my mind: Why do the artists need the money more than me because they're already super-rich? His perspective as an artist is very interesting, because one would think that he would only be supportive of artists and the RIAA. Instead, he refuses to redeem the morality of either side, only adding to the murkiness of this issue.

Ok, so there you have it: while I have a moral opposition to "pirating" music, the proposition of free music makes this opposition purely philosophical. All I have left to do is hope that "Weird Al" was wrong when in "Don't Download This Song," he calmly describes the only possible progression for a "music pirate": "You'll start out stealing songs, then you're robbing liquor stores, and selling crack, and running over school kids with your car." Oh, and also that I never get good enough at guitar to have my music "stolen" on LimeWire.[9]

I've never met the author of this essay. But I meet his kind all the time. They are my students. They populate the Stanford campus. And their attitude reveals a cost tied to the war we now wage.

No politician has explained how the benefits we seek in this war against piracy could ever justify that cost. No politician supporting this war has ever considered the alternatives to this war, and the costs. We have a practiced blindness when it comes to waging "war." We don't think about costs. We think instead about the morality of our cause. But as history has taught us again and again, morality in motive does not guarantee morality in result. Good intentions are a first step. Responsibility requires considering, and reconsidering, every step after that.

We as a people need to remember: government power is limited. It is not limited because the government has limited funds. Or limited bullets. It is limited because it operates against a background of basic morality. That morality insists upon proportionality. The parent who beats his child with a two-by-four because the child didn't clean his room is not wrong to insist his child clean his room. He is wrong because however right the motive, means are always subject to measure. A parent, an army, a government: they all must be certain that their devotion to truth does not blind them to the consequences of their actions. There's only so much a government can do. Where we find that limit, we must then find other means to the legitimate end.

CONCLUSION

The economic theory behind copyright justifies it as a tool to deal with what economists call the "problem of positive externalities."[1] An "externality" is an effect that your behavior has on someone else. If you play your music very loudly and wake your neighbors, your music is producing an externality (noise). If you renovate your house and add a line of beautiful oak trees, your renovation produces an externality (beauty). Beauty is a positive externality—people generally like to receive it. Noise is a negative externality—people (especially at 3 a.m.) don't like to receive it.

Copyright law deals with the positive externality produced by the nature of creative work. Creative work is a "public good"—meaning that (1) once it is shared, anyone can consume it without reducing the amount anyone else has; and (2) it is hard to restrict anyone from consuming it once it is available to all. If you paint a beautiful mural on your garage door, my viewing it doesn't reduce your opportunity to view it. And without building a wall around your garage (not a very practical design, for a garage at least), it's very hard to block who gets to see your mural.

Jefferson put the same idea more lyrically in a letter he wrote in 1813:

> If nature has made any one thing less susceptible than all others of exclusive property, it is the action of the thinking power called an idea, which an individual may exclusively possess as long as he keeps it to himself; but the moment it is divulged, it forces itself into the possession of every one, and the receiver cannot dispossess himself of it. Its peculiar character, too, is that no one possesses the less, because every other possesses the whole of it. He who receives an idea from me, receives instruction himself without lessening mine; as he who lights his taper at mine, receives light without darkening me. That ideas should freely spread from one to another over the globe, for the moral and mutual instruction of man, and improvement of his condition, seems to have been peculiarly and benevolently designed by nature, when she made them, like fire, expansible over all space, without lessening their density at any point, and like the air in which we breathe, move, and have our physical being, incapable of confinement or exclusive appropriation.[2]

Jefferson was talking about ideas here. Copyright regulates expression. But his observations about the nature of ideas are increasingly true of expression. If I post this book on the Internet, then your taking a copy doesn't remove my having a copy too ("no one possesses the less, because every other possess the whole"). And my making a copy available for you to have makes it relatively difficult to prevent others from having a copy as well ("like the air in which we breathe, move, and have our physical being, incapable of confinement or exclusive appropriation"). "Relatively difficult," not impossible: the whole history of Digital Rights Management technology has been the aim to remake Jefferson's nature—to make it so digital objects are like physical objects (your taking one copy means one less for me; your getting access means I don't have access).[3] But in the state of Internet nature, Internet expression is like Jefferson's ideas.

I said that economists justify copyright as a way to deal with the "problem of positive externalities." But why, you might rightly wonder, are "positive externalities" a problem? Why isn't it a positive good that expression "should freely spread from one to another over the globe, for the moral and mutual instruction of man, and improvement of his condition"? Why isn't it "peculiarly and benevolently" the Internet's "nature," to be encouraged rather than restricted?

The answer, for the economist at least, is that while free is no doubt good, if everything were free, there would be too little incentive to produce. And if there's not enough monetary incentive to produce, the economist fears, then not enough stuff is produced.

In this book I've sketched a bunch of obvious replies to this fear: there are tons of incentives beyond money. Look at the sharing economy. Look at 100 million blogs, only 13 percent of which run ads.[4] Look at Wikipedia or Free Software. Look at academics or scientists. We have plenty of examples of creative expression produced on a model different from the one that Britney Spears employs.

But I've also made the other side to that argument clear: the sharing economy notwithstanding, there's lots that won't be created without an effective copyright regime too. I love terrible Hollywood blockbusters. If anyone could copy in high quality a Hollywood film the moment it was released, no one could afford to make $100 million blockbusters. So give me this example at least. And if there's one example, then it's plausible that there are more. Movies. Maybe music. Maybe some kinds of books—dictionaries, maybe novels by John Grisham. We should of course be skeptical about how broadly this regulation needs to reach. (Supreme Court justice Stephen Breyer got tenure at Harvard with a piece that expressed deep skepticism about how broadly this claimed need reaches.)[5] But I'm convinced that it reaches into some places at least. For those cases, without solving the problem of positive externalities, we wouldn't have that kind of creative work.

So to get Hollywood films, some kinds of blockbuster movies, maybe Justin Timberlake–like music, and maybe a few types of books, we run a copyright system. That system is a form of regulation. Like most regulation, after a while, it becomes big and expensive. Federal courts and federal prosecutors spend a lot of money enforcing the law copyright is. Companies invest millions in technologies for protecting copyrighted material. Universities run sting operations on their own students to punish or expel those who fail to follow copyright's rule. We build this massively complex system of federal regulation—a regulation that purports to reach everyone who uses a computer—to solve this "problem" of positive externalities.

Good for us. Our government is working hard to "solve" this "problem." But what about negative externalities? What does our government do about those? Think, for example, about mercury spewed as pollution in the exhaust from coal-fired power plants. Or think about the carbon spewed from these coal-fired power plants. These too are externalities. Millions are exposed to dangerous levels of mercury because of this pollution. The planet teeters on a catastrophic climate tipping point because of this carbon. Whatever harm there might be in not having yet another *Star Trek*, the harms from these negative externalities are unquestionable and real. They cause real deaths. They will cause extraordinary dislocation and economic harm. So given its keen interest in regulating to protect against uncompensated positive externalities, what precisely has our government done about undoubtedly harmful negative externalities? In the past ten years, in a time when Congress has passed at least twenty-four copyright bills,[6] and federal prosecutors and federal civil courts have been used to wage "war" on "piracy" so as to solve the problem of positive externalities, what exactly has the government been doing about these negative externalities?

The answer is, not much. Though President Bush successfully deflected Al Gore's charge in 2000 that we faced a carbon crisis by promising to tax carbon when elected, within two weeks of his swearing in, he reversed himself, and indicated he didn't think global warming was a problem.[7] And though the Clean Air Act plainly regulates pollutants like mercury in power plants, in 2003, the Bush administration changed the regulations to "allow polluters to avoid actually having to reduce mercury."[8] Thus, with these real and tangible harms caused by negative externalities, the government has done worse than nothing. At the same time, it has devoted precious resources to fighting a problem that many don't even believe is a problem at all.

So what gives?

It's been a decade since I got myself into the fight against copyright extremism. Throughout this book, I have argued that this decade's work has convinced me that this war is causing great harm to our society. Not only from losses in innovation. Not only from the stifling of certain kinds of creativity. Not only because it unjustifiably limits constitutionally guaranteed freedoms. But also, and most important, because it is corrupting a whole generation of our kids. We wage war against our children, and our children will become the enemy. They will become the criminals we name them to be. And because there is no good evidence to suggest that we will win this war, that's all the reason in the world to stop these hostilities—especially when there are alternatives that advance the purported governmental interest without rendering a generation criminal.

But there is insult to add to this injury. For the point is not just that our government is waging a hopeless war. It is that our government does little to fight real harm, while it wastes resources fighting "problems" that are not even clear harms.

And why does it do this?

The lesson a decade's work has taught me is that the reason has nothing to do with stupidity. It has nothing to do with ignorance. The simple reason we wage a hopeless war against our kids is that they have less money to give to political campaigns than Hollywood does. The simple reason we do nothing while our kids are poisoned with mercury, or the environment is sent over the falls with carbon, is that our kids and our environment have less money to give to campaigns than the utilities and oil companies do. Our government is fundamentally irrational for a fundamentally rational reason: policy follows not sense, but dollars.

Until that problem is solved, a whole host of problems will go unsolved. Global warming, pollution, a skewed tax system, farm subsidies: our government is irrational because it is, in an important way, corrupt. And until that corruption is solved, we should expect little good from this government.

This book is not about that corruption generally. All I have aimed for here is to get you to take one small step. Whatever you think about global warming, the environment, tax gifts to favored corporations, subsidies that benefit only corporate farmers, at least think this: there is no justification for the copyright war that we now wage against our kids. Demand that the war stop now. And once it is over, let's get on to the hard problem of crafting a copyright system that nurtures the full range of creativity and collaboration that the Internet enables: one that builds upon the economic and creative opportunity of hybrids and remix creativity; one that decriminalizes the offense of being a teen.

ACKNOWLEDGMENTS

This book is the culmination of a long effort to make an obvious point clear. Much of that effort came in the form of lectures I've given about this subject since publishing *Free Culture* (2004). There have been more than two hundred such lectures. Each time, the argument advanced and changed, sometimes slightly, sometimes significantly. This book represents an end to that evolution, if only because I have shifted the focus of my work elsewhere.

Throughout this period, I have felt as if I was mediating between two powerful views—one expressed in the work of the late Lyman Ray Patterson; the other, expressed in the passion of the late Jack Valenti.

Patterson was a law professor at the University of Georgia. He was one of the first copyright scholars to look at the law of copyright and the reality of modern society and conclude something was profoundly out of order. Valenti was the president of the Motion Picture Association of America. He too looked at the law of copyright and the reality of modern society, and concluded something was profoundly out of order. For Patterson, the something out of order was the law. For Valenti, it was the society—or at least the youth of our society. Both worked until their dying days to correct the wrong that each had identified.

I met Patterson just once. I debated Jack Valenti publicly four times. And though the views of these two men could not have been more different, in the end I realized that the peculiar focus of this book, and of my work these past four years, was largely due to the powerful influence of both men. Were they still alive, I would have asked before I mixed them together in a single dedication. But as each devoted much of his life to teaching (even if in very different ways), I trust they both would have allowed the lesson that this particular remix might teach.

I am grateful to the many whose ideas and arguments I've used in this book and who have fundamentally shaped my thinking. I had a very different conception of the story this book tells, for example, until Tim O'Reilly shifted my view fundamentally. Likewise, though in differing degrees, with the other interviewees who appear throughout the book: Brian Behlendorf, Marc Brandon, Candice Breitz, Stewart Butterfield, Steve Chen, Gregg Gillis, Mark Hosier, Joi Ito, Mimi Ito, Don Joyce, Brewster Kahle, Heather Lawver, Declan McCullagh, Dave Marglin, Craig Newmark, Silvia Ochoa, Tim O'Reilly, Philip Rosedale, Mark Shuttleworth, Johan Söderberg, Victor Stone, Jimmy Wales, Jerry Yang, and Robert Young. I have learned a great deal from all of them, and I hope I have fairly evinced some of that understanding here.

Three other interviewees spent a great deal of time teaching me material I didn't get to use here. Dana Boyd generously shared her rich and extraordinarily interesting learning about youth and creativity. In the end, I came to believe that that research should first be presented by her. Count me among those to acknowledge it as profoundly important to an understanding of this next generation. Benjamin Mako Hill and Erik Möller spent a great deal of time outlining a rich and sophisticated understanding of "free culture." But that work complemented and corrected much of what I said in *Free Culture,* and it would have diverted the story too much here. Suffice it to say there is much more to be said, and I am hopeful I get a chance to say some of it.

I am also extraordinarily grateful to an amazing group of students who helped correct my errors and show me parts in the argument that I had missed or needed to read. They include Shireen Barday, Kevin Donovan, Paul Gowder, Jonathan Lubin, Erika Myers, and Michael Weinberg. Much of the work of coordinating these students was done by another insanely efficient and insightful "chief" research assistant, Tracy Rubin. And Christina Gagnier did a superhuman job in not only providing essential research, but pulling together that material to check everything I've said in this book. This is the second book that Christina helped make possible, and I am very thankful to her for that.

In addition to the three interviews that don't appear here, there is a massive empirical project that didn't teach enough to include. The search company Alexa contributed generously to that project by providing a listing of the top one hundred thousand sites on the Web. A massive team of volunteers from the Creative Commons community helped interpret the substance of those sites to determine the kind of interaction each encouraged. In the end, however, the results were too ambiguous to be meaningful for this book. I hope to complete the work for an essay I will publish later. But I am especially grateful to the many who helped coordinate this massive interpretive project, including Dr. Emre Bayamlioğlu, Bodó Balázs, Lu Fang, Lital Leichtag, J. C. De Martin, Dragoslava Pefeva, Jon Phillips, Song Shi, Anas Tawileh, Hung V. Tran, John Hendrik Weitzmann, and Fumi Yamazaki.

There is also a long list of friends and Internet friends who provided advice and essential information. They include Pablo Francisco Arrieta, Sean Ferry, Andy Moravcsik, and Cory Ondrejka. Dan Kahan helped me think through the effects of bad law on norms. BigChampagne (www.

bigchampagne.com) provided comprehensive data about peer-to-peer file-sharing practices. I am grateful they did so in the spirit of the "sharing economy," as the budget for a book like this could not have afforded much more.

I started writing this book at The American Academy in Berlin. I am very thankful to that bit of paradise on earth, and for its executive director, Gary Smith, who convinced me to pursue it. I am also grateful to Stanford Law School, and Dean Larry Kramer, for endless support to help me finish it.

And, as anyone who has worked with me these past few years knows, I couldn't have begun to do this and many other things without the extraordinary gift of an absolutely perfect assistant. Elaine Adolfo is not only wildly more efficient than anyone in the world; she also practices a decency and patience that is practically unknown to this world. There's no way I could thank her adequately for her help.

Finally, to my family: *Free Culture* was published just as our first child was born. I began this book just after our second was born. As anyone who has had this particular blessing knows, nothing could compare to that joy, even if too much competes with it. Endless thanks and forever love to Bettina, who has built that particular joy with me, despite the burdens this work placed right in the middle.

NOTES

Throughout this text there are references to links on the World Wide Web. As anyone who has tried to use the Web knows, these links can be highly unstable. I have tried to remedy this instability by redirecting readers to the original source through the Web site associated with this book. For each link below, you can go to remix.lessig.org and locate the original source. If the original link remains alive, you will be redirected to that link. If the original link has disappeared, you will be redirected to an appropriate reference for the material.

Preface

1. George Lakoff and Mark Johnson, *Metaphors We Live By* (Chicago: University of Chicago Press, 1980), 156–57.
2. *Ronald Allen Harmelin v. Michigan*, 501 U.S. 957 (1991).
3. Amy Harmon, "Black Hawk Download: Moving Beyond Music, Pirates Use New Tools to Turn the Net into an Illicit Video Club," *New York Times*, January 17, 2002.
4. *Metro-Goldwyn-Mayer Studios Inc. v. Grokster Ltd.*, 545 U.S. 913 (2005).

Introduction

1. All quotes from Candice Breitz taken from an in-person interview conducted August 6, 2007.
2. E-mail to Candice Breitz, May 31, 2006.
3. All quotes from Gregg Gillis taken from an interview conducted June 21, 2007, by telephone.
4. All quotes from SilviaO taken from an interview conducted February 8, 2007, by telephone.

Chapter 1. Cultures of Our Past

1. "They Ask Protection," *Washington Post,* June 6, 1906, 4.
2. E. Fulton Brylawski and Abe Goldman, *Legislative History of the 1909 Copyright Act* (South Hackensack, N. J.: Fred B. Rothman, 1976), 24.
3. Eric von Hippel, *Democratizing Innovation* (Cambridge, Mass.: MIT Press, 2005).
4. Brylawski and Goldman, *Legislative History*, 24.
5. Sousa, "The Menace of Mechanical Music," 280.
6. Ibid., 280–81 (emphasis added).
7. See Tarla Rai Peterson, "Jefferson's Yeoman Farmer as Frontier Hero: A Self Defeating Mythic Structure," *Agriculture and Human Values* 7 (1990): 9–19; Jefferson to James Madison, October 28, 1785, Merrill D. Peterson, ed., *Thomas Jefferson: Writings* (New York: Library of America, 1984), 842; Willard Sterne Randall, *Thomas Jefferson: A Life* (New York: HarperCollins, 1994), 432; Lawrence S. Kaplan, *Thomas Jefferson: Westward the Course of Empire* (Wilmington, Del.: Scholarly Resources, 1999), 27.
8. As Harvard professor Yochai Benkler describes it:

> Music in the nineteenth century was largely a relational good. It was something people did in the physical presence of each other: in the folk way through hearing, repeating, and improvising; in the middle-class way of buying sheet music and playing for guests or attending public performances; or in the upper-class way of hiring musicians. Capital was widely distributed among musicians in the form of instruments, or geographically dispersed in the hands of performance hall (and drawing room) owners. Market-based production depended on performance through presence. It provided opportunities for artists to live and perform locally, or to reach stardom in cultural centers, but without displacing the local performers.

> Yochai Benkler, *The Wealth of Networks* (New Haven, Conn.: Yale University Press, 2006), 50–51.

9. Henry Jenkins, *Convergence Culture: Where Old and New Media Collide* (New York: New York University Press, 2006).

> Cultural production occurred mostly on the grassroots level; creative skills and artistic traditions were passed down mother to daughter, father to son. Stories and songs circulated broadly, well beyond their points of origin, with little or no expectation of economic compensation; many of the best ballads or folktales come to us today with no clear marks of individual authorship. While new commercialized forms of entertainment—the minstrel shows, the circuses, the showboats—emerged in the mid-to-late nineteenth century, these professional entertainments competed with thriving local traditions of barn dances, church sings, quilting bees, and campfire stories. There was no pure boundary between the emergent commercial culture and the residual folk culture: the commercial culture raided folk culture and folk culture raided commercial culture. (Jenkins, *Convergence Culture,* 135.)

But the twentieth century, Jenkins argues, changed this:

> The story of American arts in the twentieth century might be told in terms of the displacement of folk culture by mass media. Initially, the emerging entertainment

industry made its peace with folk practices, seeing the availability of grassroots singers and musicians as a potential talent pool, incorporating community sing-a-longs into film exhibition practices, and broadcasting amateur-hour talent competitions. The new industrialized arts required huge investments and thus demanded a mass audience. The commercial entertainment industry set standards of technical perfection and professional accomplishment few grassroots performers could match. The commercial industries developed powerful infrastructures that ensured that their messages reached everyone in America who wasn't living under a rock. Increasingly, the commercial culture generated the stories, images, and sounds that mattered most to the public. (Ibid.)

10. Wikipedia contributors, "Player Piano," Wikipedia: The Free Encyclopedia, available at link #1 (last visited July 30, 2007).
11. Ibid., available at link #2 (last visited July 30, 2007).
12. Pekka Gronow, "The Record Industry: The Growth of a Mass Medium," *Popular Music* 3 (1983): 54–55, available at link #3.
13. Leonard DeGraff, "Confronting the Mass Market: Thomas Edison and the Entertainment Phonograph," *Business and Economic History* 24 (1995): 88, available at link #4.
14. Gronow, "The Record Industry," 62, available at link #5.
15. Ibid., 62, available at link #6.
16. Ibid., 63, available at link #7.
17. Philip E. Meza, *Coming Attractions? Hollywood, High Tech, and the Future of Entertainment* (Stanford, Calif.: Stanford University Press, 2007), 51.
18. Ibid., 50–51.
19. Lawrence Lessig, *Free Culture* (New York: Penguin Press, 2004), 3–7.
20. U.S. Census Bureau, "NAICS 5111—Newspaper, Periodical, Book and Directory Publishers," Industry Statistics Sampler, available at link #8 (last visited August 20, 2007).
21. U.S. Census Bureau, "NAICS 515120—Television Broadcasting," Industry Statistics Sampler, available at link #9 (last visited August 20, 2007).
22. U.S. Census Bureau, "NAICS 512—Motion Picture and Sound Recording Industries," Industry Statistics Sampler, available at link #10 (last visited August 20, 2007).
23. "Economies," Motion Picture Association of America, available at link #11 (last visited January 24, 2008).
24. Brylawski and Goldman, *Legislative History,* 25.
25. Currier was skeptical of unlimited patent rights. See *Zoltek Corp. v. United States*, 464 F.3d 1335, 1337 (2006) (Newman, J., dissenting), available at link #12 (last visited July 30, 2007). He was reluctant to expand copyrights.

Chapter 3. RO, Extended

1. Until 1972, federal law didn't regulate the copying of recordings. See *Capitol Records Inc. v. Naxos of America,* 4 N.Y.3d 540, 544 (2005).
2. See Lawrence Lessig, *Code 2.0* (New York: Basic Books, 2006), 5–6.
3. Bruce Lehman, *Intellectual Property and the National Information Infrastructure: The Report of the Working Group on Intellectual Property Rights* (Darby: Diane Publishing, 1995).

4. See Sonny Bono Copyright Term Extension Act, Pub. L. No. 105-298, 112 Stat. 2827 (1998) (extending the term of existing copyrights).

5. See, e.g., No Electronic Theft Act of 1997, Pub. L. No. 105-147, 111 Stat. 2678 ("NET Act"), amending 17 U.S.C. §506(a).

6. See Digital Millennium Copyright Act, Pub. L. No. 10-5-304, 112 Stat. 2860 (1998).

7. See *UMG Recordings Inc. v. MP3.com Inc.*, 92 F. Supp. 2d 349 (S.D.N.Y. 2000); *A&M Records Inc. v. Napster Inc.*, 239 F.3d 1004 (9th Cir. 2001).

8. See, for example, Lessig, *Free Culture,* chap. 3.

9. RIAA Watcher, "RIAA Watch," RIAA Watch, available at link #13; Nate Mook, "RIAA Sues 261, Including 12-Year-Old Girl," *BetaNews,* September 9, 2003, available at link #14; Nate Mook, "RIAA Sues Deceased Grandmother," *BetaNews,* February 4, 2005, available at link #15.

10. See Susan Butler, "Sixth Wave of RIAA Pre-Litigation Letters Sent to Colleges," *Hollywood Reporter,* July 19, 2007, available at link #16.

11. Number of Lawsuits in the United States Courts Concerning Copyright, 2000–2006.

Year	Court of Appeals	District Courts
2006	415	5,488
2005	453	4,595
2004	475	2,653
2003	453	2,111
2002	517	2,439
2001	486	2,049
2000	Not Available	2,056

The Federal Judiciary, "Federal Judicial Caseload Statistics," U.S. Courts, available at link #17 (last visited July 30, 2007). Table C-2. "U.S. District Courts—Civil Cases Commenced, by Basis of Jurisdiction and Nature of Suit," 2000–2006. Table B-7. "U.S. Courts of Appeals—Nature of Suit or Offense in Cases Arising from the U.S. District Courts, by Circuit," 2000–2006.

12. International Federation of the Phonographic Industry, *IFPI:07 Digital Music Report* (London: IFPI, 2007), 18.

13. The last great panic surrounded the emergence of cassette-tape technology. See the discussion in Office of Technology Assessment, *Copyright and Home Copying: Technology Challenges the Law* (Washington, D.C.: U.S. Government Printing Office, 1989), 145–47, available at link #18.

14. Meza, *Coming Attractions?,* 87–88.

15. Ibid. But obviously the claims are contested. See Martin Peitz and Patrick Waelbroeck, "The Effect of Internet Piracy on CD Sales: Cross-Section Evidence," *CESifo Working Paper Series* 1122 (January 2004), concluding that Internet piracy accounts for just 22.5 percent of the drop in CD sales. See also Felix Oberholzer and Koleman Strumpf, "The Effect of File-Sharing on Record Sales: An Empirical Analysis," University of North Carolina (2004) (no statistically significant connection between downloading and drop

in sales). Compare Stan J. Liebowitz, "Economists' Topsy-Turvy View of Piracy," *Review of Economic Research on Copyright Issues 2* (2005): 5–17; Stan J. Liebowitz, "Economists Examine File-Sharing and Music Sales," *Industrial Organization and the Digital Economy* (Cambridge, Mass.: MIT Press, 2005); Stan J. Liebowitz, "Testing File-Sharing's Impact on Music Album Sales in Cities," *Management Science,* forthcoming.

16. "The Recording Industry 2006 Piracy Report: Protecting Creativity in Music," International Federation of the Phonographic Industry (IFPI), available at link #19 (last visited January 18, 2008). For a fantastic study of the relationship between artists' income and digital technologies see Martin Kretschmer, "Artists' Earning and Copyright: A Review of British and German Music Industry Data in the Context of Digital Technologies," *First Monday* 10 (2005), available at link #20.

17. Amy Matthew, "The Creative Revolution: Consumers, Artists Lead the Way into New Entertainment World," *Pueblo Chieftain,* April 29, 2007; "Apple to Give iTunes Users Credit for Full Albums," *InvesTrend*, March 29, 2007.

18. Less elegant was the idea of a coin box in the home. I remember as a kid staying in a boarding house in England in the 1970s that had a coin box to operate hot water. A hot bath is one thing, but television is a necessity. The inconvenience of this system must have been staggering. Still, the idea was tested in Palm Springs, California, in the early 1950s, again by Paramount. This system, called telemeter, used scrambled television signals sent over telephone lines. When customers deposited the correct amount of change in the coin box, the telemeter descrambled the signal. The system debuted in 1953, showing a USC–Notre Dame football game (presaging the popular college sports packages now available on satellite TV services) for $1.00 and a first-run Paramount movie, *Forever Female,* for an additional $1.35. (Meza, *Coming Attractions?,* 83.)

19. In Napster Inc. Copyright Litigation, 191 F. Supp. 2d 1087 (N.D. Cal. 2002).

20. In January 2008 the last of the major labels, Sony BMG, dropped the requirement that its music be sold with DRM. Peter Sayer, "Sony BMG to Sell DRM-Free Music Downloads Through Stores," *InfoWorld,* January 7, 2008, available at link #21.

21. Lessig, *Code 2.0.*

Chapter 4. RW, Revived

1. *Grand Upright Music Ltd. v. Warner Bros. Records Inc.,* 780 F. Supp. 182 (S.D.N.Y. 1991).

2. All quotes from Don Joyce taken from an interview conducted March 20, 2007, by telephone.

3. David Bollier, *Brand Name Bullies* (Hoboken, N.J.: Wiley, 2005), 69.

4. J. D. Lasica, *Darknet: Hollywood's War Against the Digital Generation* (Hoboken, N.J.: Wiley, 2005), 72–73.

5. Heidi Anderson, "Plugged In," *Smart Computing* (November 2000): 90–92.

6. Benkler, *Wealth of Networks,* 217.

7. Mark Lawson, "Berners-Lee on the Read/Write Web," *BBC News*, August 9, 2005, available at link #22 (last visited July 31, 2007).

8. Niall Kennedy, "Technorati Two Years Later," Niall Kennedy's Weblog, November 26, 2004, available at link #23; David Sifry, "Technorati," Sifry's Alerts, November 27, 2002, available at link #24; David Sifry, "Over 100,000 Blogs Served," Sifry's Alerts, March 5, 2003, available at link #25; David Sifry, "One Million Weblogs Tracked," Sifry's

Alerts, September 27, 2003, available at link #26; David Sifry, "State of the Blogosphere," Sifry's Alerts, October 10, 2004, available at link #27; "About Us," Technorati, available at link #28 (last visited July 30, 2007).

9. David Sifry, "The State of the Live Web," Sifry's Alerts, April 5, 2007, available at link #29 (last visited July 23, 2007); David Sifry, "State of the Blogosphere, October 2006," Technorati, available at link #30 (last visited July 23, 2007).

10. Benkler, *Wealth of Networks,* 217.

11. Thomas Vander Wal, "Off the Top: Folksonomy Entries," Vanderwal.net, October 3, 2004, available at link #31.

12. Don Tapscott and Anthony D. Williams, *Wikinomics: How Mass Collaboration Changes Everything* (New York: Portfolio, 2006), 41.

13. Ibid., 52.

14. Ibid., 144–45.

15. Ibid., 42.

16. "Blogging Basics," Technorati, available at link #32 (last visited July 23, 2007).

17. David Sifry, "The State of the Live Web, April 2007," Sifry's Alerts, available at link #33 (last visited August 16, 2007).

18. Charlene Li, *Social Technographics* (Cambridge, Mass.: Forrester, 2007), 2.

19. Benkler, *Wealth of Networks,* 225–33.

20. The blog Corporate Influence in the Media tries to document examples of advertisers pressuring publishers and broadcasters. See Anup Shah, Corporate Influence in the Media, available at link #34 (last visited August 16, 2007). In 2006 the Center for Media and Democracy released a report detailing a large number of instances in which news organizations broadcast "video news releases" as news without revealing to their audiences that the video was provided by a public relations firm. See Diane Farsetta and Daniel Price, "Fake TV News: Widespread and Undisclosed," Center for Media and Democracy, April 6, 2006, available at link #35.

21. Lawrence Lessig, "The People Own Ideas," *Technology Review* (June 2005): 46–48.

22. "To the Point; Trends & Innovations," *Investor's Business Daily*, September 26, 2006.

23. Ibid.

24. Gail Koch, "'800-Pound Gorilla' Still Rules for Most," *Star Press* (Muncie, Ind.), September 28, 2005.

25. Xinhua News Agency, "Census Bureau: Americans to Spend More Time on Media Next Year," December 15, 2006.

26. U.S. Fed News, "American Time Use Survey—2006 Results," *U.S. Fed News*, June 28, 2007.

27. This is not to say that before the Internet, there was nothing like this RW-media culture. Indeed, for almost a half century, beginning with the *Star Trek* series, there has been a rich "fan fiction" culture, in which fans take popular culture and remix it. Rebecca Tushnet, "Legal Fictions: Copyright, Fan Fiction, and a New Common Law," *Loyola of Los Angeles Entertainment Law Journal* 17 (1997): 655, citing Henry Jenkins and John Tulloch, eds., *'At Other Times, Like Females': Gender and* Star Trek *Fan Fiction,* in *Science Fiction Audiences: Watching* Dr. Who *and* Star Trek (London: Routledge, 1995), 196. Some trace the history of fan fiction back even earlier, to "metanovels" written in response to classic works of fiction such as *Pride and Prejudice* (Sharon Cumberland, "Private Uses of Cyberspace: Women, Desire and Fan Culture," in *Rethinking Media Change: The Aesthetics of Transition,* ed. David Thorburn and Henry Jenkins (Cambridge, Mass.: MIT

Press, 2003), 261. An Internet commentator known as Super Cat argues that the first fan fiction was John Lydgate's *The Siege of Thebes,* a continuation of *The Canterbury Tales* circa 1421: Super Cat, "A (Very) Brief History of Fanfic," Fanfic Symposium, available at link #36 (last visited August 11, 2007). Many believe that the contemporary online fan fiction community is predominantly composed of women, and the genre addresses topics traditionally marginalized in the commercial media, including "the status of women in society, women's ability to express desire, [and] the blurring of stereotyped gender lines" (Cumberland, "Private Uses," 265). In addition to traditional textual fan fiction, cyberspace has spawned an active culture of fan filmmaking. See Henry Jenkins, "Quentin Tarantino's Star Wars? Digital Cinema, Media Convergence, and Participatory Culture," in *Rethinking Media Change: The Aesthetics of Transition,* ed. David Thorburn and Henry Jenkins (Cambridge, Mass.: MIT Press, 2003), 281–84 (offering a case study of *Star Wars* fan fiction, which began in textual form with the first official film, and developed into digital film distributed on independent creators' Web sites). There is a comprehensive study of fan fiction in chapters 5–8 of Henry Jenkins, *Textual Poachers: Television Fans and Participatory Culture* (New York: Routledge, 1992).

28. All quotes from Mark Hosler taken from an interview conducted May 1, 2007, by telephone.

29. All quotes from Johan Söderberg taken from an interview conducted February 15, 2007, by telephone.

30. Telephone interview with Don Joyce, March 20, 2007.

31. "Misperceptions, the Media, and the Iraq War," Program on International Policy Attitudes and Knowledge Networks, available at link #37 (last visited January 18, 2008).

32. Charles Krauthammer, "A Vacation Bush Deserves," *Washington Post,* August 10, 2001.

33. All quotes from Victor Stone taken from an interview conducted February 15, 2007, by telephone.

34. All quotes from Mimi Ito taken from an interview conducted January 24, 2007, by telephone. For more, see Mimi Ito, "Japanese Media Mixes and Amateur Cultural Exchange," in *Digital Generations,* ed. David Buckingham and Rebekah Willett (Mahwah, N.J.: Lawrence Erlbaum, 2006), 49–66.

35. Jenkins, *Convergence Culture,* 128.

36. Ibid., 182.

37. Ibid., 177.

38. Ibid., 182.

Chapter 5. Cultures Compared

1. An argument "in favor" is certainly not an argument anyone should consider conclusive. Free speech values should still weigh in the balance, driving regulation away from restrictive measures when alternative, nonrestrictive alternatives exist.

2. Andrew Odlyzko, "Content Is Not King," *First Monday* 6 (2001), available at link #38.

3. Stewart Baker, "Exclusionary Rules," *Wall Street Journal,* March 26, 2004.

4. Andrew Keen, *The Cult of the Amateur* (New York: Doubleday, 2007), 64.

5. Ibid., 27.

6. Ibid., 131.

7. Ibid., 15.

306   NOTES

8. I've enumerated some errors on my blog. See Lawrence Lessig, "Keen's 'The Cult of the Amateur': BRILLIANT!" Lessig Blog, available at link #39.

9. Keen, *The Cult of the Amateur,* 27.

10. New York Institute for the Humanities and NYU Humanities Council, "The Comedies of Fair U$e," Internet Archive, available at link #40 (last visited July 30, 2007); Joy Garnett, "Full Program Audio on Archive.org," Comedies of Fair U$e, available at link #41 (last visited July 30, 2007).

11. Steven Johnson, *Everything Bad Is Good for You: How Today's Popular Culture Is Actually Making Us Smarter* (New York: Riverhead, 2005).

12. Jenkins, *Convergence Culture,* 103–4.

13. A real problem for readers of his last novel, *The Mystery of Edwin Drood* (1870). Dickens died before he completed the story, even though serial chapters were already being printed. Joel J. Brattin, "Dickens and Serial Publication," PBS, available at link #42 (last visited August 16, 2007).

14. Christopher Lydon, "Ecstasy of Influence—Interview with Jonathan Lethem, Siva Vaidhyanathan, Mark Hosler, and Mike Doughty," Open Source with Christopher Lydon, February 2, 2007, available at link # 43.

15. Ithiel de Sola Pool, *Technologies Without Boundaries: On Telecommunications in a Global Age* (Cambridge, Mass.: Harvard University Press, 1990), 121.

16. The standard of intermediate First Amendment review permits speech regulation only "[1] if it advances important governmental interests unrelated to the suppression of free speech and [2] does not burden substantially more speech than necessary to further those interests." *Turner Broad. Sys. v. FCC,* 520 U.S. 180, 189 (1997); see also *United States v. O'Brien,* 391 U.S. 367, 377 (1968); *Ward v. Rock Against Racism,* 491 U.S. 781, 791 (1989) (applying intermediate scrutiny to time, place, and manner regulation of speech in the public forum); *San Francisco Arts & Athletics Inc. v. U.S. Olympic Comm.,* 483 U.S. 522, 537 (1987) (applying *O'Brien* review to a law protecting the word "Olympic" under trademark law).

17. Work from 1923 on is potentially subject to copyright. Whether in fact a particular work is copyrighted depends upon whether the work satisfied certain formalities.

18. Jessica Litman, "The Exclusive Right to Read," *Cardozo Arts and Entertainment Law Journal* 13 (1994): 29, 34–35.

19. R. Anthony Reese, "Innocent Infringement in U.S. Copyright Law: A History," *Columbia Journal of Law & the Arts* 30 (2007): 133, 136.

20. V. Clapp, *Copyright—A Librarian's View, Prepared for the National Advisory Commission on Libraries* (Washington D.C.: Copyright Committee, Association of Research Libraries, 1968).

21. This important though obscure story about the unintended expansion of the scope of copyright is told best by L. Ray Patterson, "Free Speech, Copyright, and Fair Use," *Vanderbilt Law Review* 40 (1987): 40–43.

22. Paul Goldstein, *Copyright's Highway: From Gutenberg to the Celestial Jukebox* (Stanford, Calif.: Stanford University Press, 2003).

23. Office of Technology Assessment, *Copyright and Home Copying: Technology Challenges the Law* (Washington, D.C.: US Government Printing Office, 1989), 145–47, available at link #44.

24. Wikipedia contributors, "Jazz," Wikipedia: The Free Encyclopedia, available at link #45 (last visited July 30, 2007).

25. Wikipedia contributors, "Louis Armstrong," Wikipedia: The Free Encyclopedia, available at link #46 (last visited July 30, 2007).

26. Fairly relaxed, not completely. There is an important tension in jazz created by the way the derivative right functions. Because jazz is in essence improvisation, it must build upon some other work. But because copyright law treats this other work as expression, rather than as an idea, the improvisation requires permission from the underlying copyright owner. In practice, jazz musicians almost never seek that permission, instead relying upon the mechanical license to secure permission to record the underlying work. That license, however, doesn't cover derivatives. For a penetrating analysis of these questions, see Anonymous, "Jazz Has Got Copyright Law & That Ain't Good," *Harvard Law Review* 118 (2005), 1940.

27. *Bridgeport Music Inc. v. Dimension Films,* 383 F.3d 390 (6th Cir. 2004).

28. William W. Fisher, *Promises to Keep* (Stanford, Calif.: Stanford University Press, 2004).

29. Peter Lauria, "File-$haring," *New York Post,* June 25, 2007.

30. *Metro-Goldwyn-Mayer Studios Inc. v. Grokster Ltd.,* 545 U.S. 913 (2005).

31. Mitch Bainwol and Cary Sherman, "Explaining the Crackdown on Student Downloading," *Inside Higher Ed,* March 15, 2007, available at link #47.

32. I also stand by my view that the harms caused by p2p file sharing are overstated by the industry. Mark Cooper has now added to this debate. As he has argued effectively, much of the loss in sales comes from people buying one or two tracks from an album. LPs forced those tracks to be bundled before; digital technology now permits them to be separate. It makes no sense to count that "loss" as a harm to society, since it simply represents people choosing to buy what they want. See Mark Cooper, *Digital Downloading of Music* (Washington, D.C.: Consumer Federation of America, 2007).

33. Bainwol and Sherman, "Explaining the Crackdown."

Chapter 6. Two Economies: Commercial and Sharing

1. Yochai Benkler, "Sharing Nicely: On Shareable Goods and the Emergence of Sharing as Modality of Economic Production," *Yale Law Journal* 114 (2004): 273-358.

2. Ronald E. Yates, "Internet-Related Manager Tops List of Hottest Jobs; Position Is So New and in Such Demand That Candidates' Lack of Degrees or Advanced Age Are Not Seen as Deterrents," *Sun-Sentinel,* February 5, 1996. Robert D. Atkinson and Daniel K. Correa, "The Digital Economy—Internet Domain Names," *The 2007 State New Economy Index* (2007): 40, available at link #48.

3. U.S. Census Bureau, "E-Stats-Measuring the Electronic Economy," available at link #49.

4. *Sony Corp. v. Universal City Studios,* 464 U.S. 417 (1984).

5. See Julie Niederhoff, "Video Rental Developments and the Supply Chain: Netflix, Inc.," Washington University, St. Louis (2002); Michael K. Mills and Jon Silver, "Analysing the Effect of Digital Technology on Channel Strategy, Power and Disintermediation in the Home Video Market: The Demise of the Video Store?" *Video Technology* magazine (February 2005); IRS, "Retail Industry ATG—Chapter 3: Examination Techniques for Specific Industries (Video/DVD Rental Business)," Small Business and Self-Employed One-Stop Resource, August 2005, available at link #50; "Videotape Rental—Background and Development," All Business, available at link #51; "Videotape Rental—Current Conditions," All Business, available at link #52.

6. Hoover's Inc., "Blockbuster, Inc.," Answers.com, available at link #53 (last visited August 7, 2007).

7. Wikipedia contributors, "Blockbuster, Inc." Wikipedia: The Free Encyclopedia, available at link #54 (last visited July 31, 2007).

8. Keith Regan, "Netflix Taking Over Wal-Mart's Online DVD Rental Business," *E-Commerce Times*, May 19, 2005, available at link #55 (last visited July 5, 2007); see also Phillip Torrone, "Netflix, Open Up or Die . . . ," *Engadget*, July 19, 2004, available at link #56.

9. Hoover's Inc., "Amazon.com," Answers.com, available at link #57 (last visited July 31, 2007). These numbers reflect sales only. According to reports, Amazon's net deficit is still high—$2 billion as of 2005.

10. Ibid., available at link #58 (last visited July 31, 2007).

11. Wikipedia contributors, "Larry Page," Wikipedia: The Free Encyclopedia, available at link #59 (last visited July 31, 2007).

12. Verne Kopytoff, "Google Shares Top $400: Search Engine No. 3 in Market Cap Among Firms in Bay Area," *San Francisco Chronicle*, November 18, 2005; Yahoo! Finance, "GOOG: Key Statistics for Google Inc," Capital IQ, available at link #60 (last visited July 5, 2007).

13. Keen, *The Cult of the Amateur*, 135.

14. The point was made long before by Nicholas Negroponte. "A best-seller in 1990, Nicholas Negroponte's *Being Digital* drew a sharp contrast between 'passive old media' and 'interactive new media,' predicting the collapse of broadcast networks in favor of an era of narrowcasting and niche media on demand: 'What will happen to broadcast television over the next five years is so phenomenal that it's difficult to comprehend.'" Jenkins, *Convergence Culture*, 5.

15. Chris Anderson, *The Long Tail* (New York: Hyperion, 2006), 23.

16. Alan Cohen, "The Great Race; No Startup Has Cashed In on the DVD's Rapid Growth More Than Netflix. Now Blockbuster and Wal-Mart Want In. Can It Outrun Its Big Rivals?," *Fortune Small Business*, December 2002–January 2003. "Media Center," Netflix, available at link #61 (last visited April 1, 2008).

17. See Erik Brynjolfsson, Yu Jeffrey Hu, and Duncan Simester, "Goodbye Pareto Principle, Hello Long Tail: The Effect of Search Costs on the Concentration of Product Sales," MIT Center for Digital Business Working Paper (2007); Paul L. Caron, "The Long Tail of Legal Scholarship," *Yale Law Journal* 116 Pocket Part 38 (2006); Anita Elberse and Felix Oberholzer-Gee, "Superstars and Underdogs: An Examination of the Long Tail Phenomenon in Video Sales," Harvard Business School No. 07-015 Working Paper Series; Indiana Resource Sharing Task Force, "Wagging the Long Tail: Sharing More of Less; Recommendations for Enhancing Resource Sharing in Indiana," White Paper (2007); Anindya Ghose and Bin Gu, "Search Costs, Demand Structure and Long Tail in Electronic Markets: Theory and Evidence," NET Institute Working Paper No. 06-19 (2006); Teruyasu Murakami, "The Long Tail and the Lofty Head of Video Content: The Possibilities of 'Convergent Broadcasting,'" Nomura Research Institute, NRI Papers No. 113 (2007).

18. Lee Gomes, "It May Be a Long Time Before the Long Tail Is Wagging the Web," *Wall Street Journal*, July 26, 2006.

19. All quotes from Robert Young taken from an interview conducted April 26, 2007, by telephone.

NOTES

309

20. Free as in free speech is different. Robert Young has been a strong supporter of Creative Commons.
21. I am grateful to Tim O'Reilly for getting me to see the importance of this point.
22. Dan Bricklin, "The Cornucopia of the Commons: How to Get Volunteer Labor," Dan Bricklin's Web site, August 7, 2000, available at link #62.
23. Linked from Bricklin, "Cornucopia of the Commons."
24. Dan Bricklin, "Cornucopia of the Commons," available at link #63.
25. See "Google Defies US Over Search Data," BBC News, January 20, 2006, available at link #64; Maryclaire Dale, "Judge Throws Out Internet Blocking Law: Ruling States Parents Must Protect Children Through Less Restrictive Means," MSNBC, March 22, 2007, available at link #65. Google prevailed in its effort to restrict the government's search. See *Gonzales v. Google,* 234 F.R.D. 674 (N.D. Cal. 2006).
26. Phillip Torrone, "Netflix, Open Up or Die . . . ," available at link #66.
27. Netflix, *Netflix Prize,* available at link #67 (last visited July 2, 2007).
28. Tapscott and Williams, *Wikinomics,* 183.
29. See Tim O'Reilly, "What Is Web 2.0: Design Patterns and Business Models for the Next Generation of Software," O'Reilly, September 30, 2005, available at link #68. As Mary Madden summarizes the idea, it is "utilizing collective intelligence, providing network-enabled interactive services, giving users control over their own data." Mary Madden and Susannah Fox, *Riding the Waves of Web 2.0* (Washington, D.C.: Pew Internet Project, 2006), 1.
30. Ronald H. Coase, "The Nature of the Firm," *Economica* 4 (1937): 386–405.
31. Benkler, *The Wealth of Networks,* 59–60.
32. Lawrence Lessig, *The Future of Ideas* (New York: Random House, 2001) 35–36.
33. Clayton M. Christensen, *The Innovator's Dilemma* (Boston, Mass.: Harvard Business School Press, 1997), 228.
34. Benkler, "Sharing Nicely," 282.
35. This is the phenomenon of "crowding out" described extensively by Professor Benkler in *The Wealth of Networks.* As he summarizes this work, "Across many different settings, researchers have found substantial evidence that under some circumstances, adding money for an activity previously undertaken without price compensation reduces, rather than increases, the level of activity" (94).
36. See Michael Walzer, *Spheres of Justice: A Defense of Pluralism and Equality* (New York: Basic Books, 1984).
37. Lewis Hyde, *The Gift—Imagination and the Erotic Life of Property* (New York: Vintage Books, 2004), 3.
38. Ibid., 56.
39. Ibid., 45–46.
40. Ibid., 82.
41. Benkler, "Sharing Nicely," 327.
42. Ibid.
43. Ibid., 324; see also at 323, describing the work of Bruno Frey.
44. Increasingly the concern among record company executives is with social sharing. See Jason Pontin, "A Social-Networking Service with a Velvet Rope," *New York Times,* July 29, 2007.

45. See Eric A. von Hippel and Karim Lakhani, "How Open Source Software Works: 'Free' User-to-User Assistance," *Research Policy* 32 (2003): 923–43.

Kollock (1999) discusses four possible motivations to contribute public goods online. Given that his focus is incentives to put online something that has already been created, his list does not include any direct benefit from developing the thing itself—either the use value or the joy of creating the work product. His list of motives to contribute does include the beneficial effect of enhancements to one's reputation. A second potential motivator he sees is expectations of reciprocity. Both specific and generalized reciprocity can reward providing something of value to another. When information providers do not know each other, as is often the case for participants in open source software projects, the kind of reciprocity that is relevant is called "generalized" exchange (Ekeh, 1974). . . . The third motivator posited by Kollock is that the act of contributing can have a positive effect on contributors' sense of "efficacy"—a sense that they have some effect on the environment (Bandura, 1995). Fourth and finally, he notes that contributors may be motivated by their attachment or commitment to a particular open source project or group. In other words, the good of the group enters into the utility equation of the individual contributor. (Ibid., 927.)

46. Or so the terms of service for Skype say. See "Skype End User License Agreement—Article 4 Utilization of Your Computer," Skype, available at link #69 (last visited July 31, 2007).
47. Daniel H. Pink, "The Book Stops Here," *Wired,* March 2005, available at link #70.
48. All quotes from Jimmy Wales taken from an in-person interview conducted May 4, 2007.
49. Seth Anthony, "Contribution Patterns Among Active Wikipedians: Finding and Keeping Content Creators," Wikimania Proceedings SA1 (2006), as summarized at link #71 (last visited August 20, 2007).
50. Aaron Swartz, "Who Writes Wikipedia," available at link #72 (last visited August 20, 2007).
51. "Meetings/February 7, 2005," Wikimedia Foundation, available at link #73 (last visited July 31, 2007).
52. Tapscott and Williams, *Wikinomics,* 72.
53. Ibid.
54. Noam Cohen, "The Latest on Virginia Tech, from Wikipedia," *New York Times,* April 23, 2007.
55. Ibid.
56. Robert Young and Wendy Goldman Rohm, *Under the Radar: How Red Hat Changed the Software Business—and Took Microsoft by Surprise* (Scottsdale, Ariz.: Coriolis Group Books, 1999), 110.
57. Netcraft, "Reports—What Is the Market Share of the Different Servers?" Netcraft—Web Server Survey, available at link #74 (last visited July 31, 2007): follow monthly "Index" link for November 1996–present; follow monthly "ALL" link for August 1995–October 1996.
58. Steven Weber, *The Success of Open Source* (Cambridge, Mass.: Harvard University Press, 2004), 234.
59. Scott E. Page, *The Difference: How the Power of Diversity Creates Better Groups, Firms, Schools, and Societies* (Princeton, N.J.: Princeton University Press, 2007).
60. Wikipedia contributors, "Project Gutenberg," Wikipedia: The Free Encyclopedia, available at link #75 (last visited October 10, 2007).

61. Ibid., available at link #76 (last visited October 10, 2007).

62. "Beginning Proofreaders' Frequently Asked Questions," Distributed Proofreaders, available at link #77 (last visited July 31, 2007).

63. Wikpedia contributors, "SETI@home," Wikipedia: The Free Encyclopedia, available at link #78 (last visited August 20, 2007). See also Benkler, "Sharing Nicely," 275.

64. Wikpedia contributors, "Einstein@Home," Wikipedia: The Free Encyclopedia, available at link #79 (last visited August 20, 2007).

65. "About the Internet Archive," Internet Archive, available at link #80 (last visited July 31, 2007).

66. All quotes from Brewster Kahle taken from an interview conducted January 24, 2007, by telephone.

67. NASA Ames, "Welcome to the Clickworkers Study," Clickworkers, available at link #81 (last visited July 31, 2007).

68. B. Kanefsky, N. G. Barlow, and V. C. Gulick, "Can Distributed Volunteers Accomplish Massive Data Analysis Tasks," Thirty-second Annual Lunar and Planetary Science Conference 1272 (2001), available at link #82.

69. Ibid., available at link #83.

70. Ibid., available at link #84.

71. Benkler, The Wealth of Networks, 69.

72. "Let Data Speak to Data," Nature 438 (2005), available at link #85 (last visited July 31, 2007), cited in Tapscott and Williams, Wikinomics, 159.

73. Ibid., available at link #86 (last visited July 31, 2007).

74. Michael W. Vannier and Ronald M. Summers, "Sharing Images," Radiology 228 (2003), available at link #87.

75. U.S. National Virtual Observatory, available at link #88 (last visited July 31, 2007).

76. "About the Open Directory Project," Open Directory Project, available at link #89 (last visited July 11, 2007).

77. "About Open Source Food," Open Source Food, available at link #90 (last visited July 11, 2007).

78. Benkler, The Wealth of Networks, 121.

79. Bricklin, "The Cornucopia of the Commons," available at link #91.

80. von Hippel, Democratizing Innovation, 60–61.

81. Steven Weber, The Success of Open Source (Cambridge, Mass.: Harvard University Press, 2004), 153.

82. von Hippel, Democratizing Innovation, 60–61.

83. Weber, The Success of Open Source, 155.

84. von Hippel and Lakhani, "How Open Source Software Works," 927.

85. Weber, The Success of Open Source, 224.

86. Benkler, The Wealth of Networks, 16–18.

87. Ibid., 2.

88. Ibid., 3.

Chapter 7. Hybrid Economies

1. I don't mean to suggest that there weren't hybrids before the Internet. Think of dog shows, open-mike nights at bars, or country fairs. All of these have a dynamic similar to

the one I identify on the Internet. The only difference is the significance of these hybrids. The Internet will enable a much wider range of hybridization, with a much greater economic and social value. I am grateful to Oliver Baker for reminding me of this point.

2. Benkler, *The Wealth of Networks,* 55.

3. "GNU Free Documentation License," Free Software Foundation, available at link #92 (last visited August 20, 2007).

4. All quotes from Brian Behlendorf taken from an interview conducted May 11, 2007, by telephone.

5. Red Hat employed 50 percent of Linux's core team. Telephone interview with Robert Young, April 26, 2007.

6. Red Hat and VA Linux gave stock options to Torvalds. Wikipedia contributors, "Linus Torvalds," Wikipedia: The Free Encyclopedia, available at link #93 (Last visited July 31, 2007); Gary Rivlin, "Leader of the Free World," *Wired,* November 2003, available at link #94.

7. All quotes from Mark Shuttleworth taken from an interview conducted March 19, 2007, by telephone.

8. Ted Rheingold, "Don't Outsource Your Sales," Dogster & Catster company Blogster, available at link #95 (last visited April 1, 2008).

9. All quotes from Internet venture capitalist Joichi Ito taken from an interview conducted January 23, 2007, by telephone.

10. "New Cyberspace Classified Ad Technologies to Impact Newspaper Revenues in 3 Years," *PR Newswire,* November 21, 1996.

11. "Newspaper Advertising Being Challenged by the 'Net," *E-Commerce Law Report* 11 (1999), 24.

12. Wikipedia contributors, "craigslist," Wikipedia: The Free Encyclopedia, available at link #96 (last visited July 31, 2007).

13. Ibid., available at link #97 (last visited July 31, 2007).

14. All quotes from Craig Newmark taken from an interview conducted January 22, 2007, by telephone.

15. Tapscott and Williams, *Wikinomics,* 190. craigslist charges for job ads.

16. Ibid.

17. Ibid.

18. Nick C. Sortal and Ian Katz, "Volunteers Use Internet to Offer Homes to Katrina Victims," *South Florida Sun-Sentinel,* September 1, 2005.

19. Kathleen Sullivan, "Hurricane Katrina; Lodging Offers Being Posted on Craigslist; Across Continent, People Opening Homes to Survivors," *San Francisco Chronicle,* September 4, 2005.

20. Tapscott and Williams, *Wikinomics,* 187.

21. "Flickr Founder: Creativity Is Human Nature," CNN.com, January 17, 2007, available at link #98; Dan Fost, "Welcoming Startups into Yahoo's Fold; Web Portal Works to Integrate the Companies It Has Acquired," *San Francisco Chronicle,* December 24, 2006, available at link #99.

22. All quotes from Stewart Butterfield taken from an interview conducted May 1, 2007, by telephone.

23. All quotes from Steve Chen taken from an interview conducted January 29, 2007, by telephone.

24. Wikipedia contributors, "YouTube," Wikipedia: The Free Encyclopedia, available at link #100 (last visited July 31, 2007); Pete Cashmore, "YouTube Is World's Fastest Growing Website," *Mashable—Social Networking News,* July 22, 2006, available at link #101.

25. Wikipedia contributors, "List of YouTube Celebrities," Wikipedia: The Free Encyclopedia, available at link #102 (last visited April 1, 2008).

26. Tim Deal, *User-Generated Video on the Web: A Taxonomy and Market Outlook* (Silver Spring, Md.: Pike & Fischer, 2007), 4.

27. All quotes from Declan McCullagh taken from an interview conducted April 4, 2007, by telephone.

28. Declan McCullagh, "About Declan McCullagh's Politech," Politech: Politics & Technology, available at link #103 (last visited July 31, 2007).

29. "Yahoo! Groups," Yahoo!, available at link #104 (last visited January 18, 2008).

30. Tapscott and Williams, *Wikinomics,* 259.

31. All quotes from Jerry Yang taken from an interview conducted January 24, 2007, by telephone.

32. "Yahoo! Answers," Yahoo!, available at link #105 (last visited April 14, 2008).

33. "Points and Levels," Yahoo! Answers Point System, available at link #106 (last visited August 20, 2007).

34. "Bessemer Venture Partners Funds Jimmy Wales' Startup Wikia," Wikia.com, available at link #107 (last visited July 31, 2007).

35. Michael Arrington, "Wikia Gaming Launches with 250,000 Articles," TechCrunch, available at link #108 (last visited January 18, 2008).

36. Jenkins, *Convergence Culture,* 185.

37. Ibid.

38. All quotes from Marc Brandon taken from an interview conducted February 13, 2007, by telephone.

39. Jenkins, *Convergence Culture,* 185.

40. All quotes from Heather Lawver taken from an interview conducted February 1, 2007, by telephone.

41. Telephone interview with Heather Lawver, February 1, 2007.

42. Jenkins, *Convergence Culture,* 187.

43. Elizabeth Weise, "'Potter' Fans Put Hex of a Boycott on Warner Bros.," *USA Today,* February 22, 2001; Janine A. Zeitlin, "Potter Fan Masters Her Domain," *Washington Times,* July 19, 2001.

44. Jenkins, *Convergence Culture,* 187.

45. Ibid., 186.

46. Ibid., 58.

47. Tapscott and Williams, *Wikinomics,* 135–36.

48. Jenkins, *Convergence Culture,* 3.

49. Ibid., 96.

50. Ibid., 133.

51. Maria Alena Fernandez, "ABC's *Lost* Is Easy to Find, and Not Just on a TV Screen," *Los Angeles Times,* January 3, 2006.

52. Ibid.

53. Ibid.

54. Tapscott and Williams, *Wikinomics,* 127.

55. All quotes from Philip Rosedale taken from an interview conducted February 8, 2007, by telephone.

56. Wagner James Au, "Laying Down the Law: The Notary Public of Thyris, New World Notes," New World Notes, available at link #109 (last visited July 31, 2007).

57. "State of Play III—Social Revolutions," State of Play Architecture Submissions, available at link #110 (last visited August 20, 2007).

58. Sherry Turkle, *Life on the Screen: Identity in the Age of the Internet* (New York: Simon & Schuster, 1995), 13.

59. Tapscott and Williams, *Wikinomics,* 30.

60. Ibid., 39.

61. Ibid., 45.

62. All quotes from Tim O'Reilly taken from an interview conducted January 24, 2007, by telephone.

Chapter 8. Economy Lessons

1. See Ed Treleven, "Cracking Down of Music Theft; Recording Industry Gets Very Aggressive," *Wisconsin State Journal,* February 4, 2007. ("Big Champagne found that in August 2003, when the RIAA began bringing lawsuits against file-sharers, the average number of global peer-to-peer users online at one time was about 3.8 million. Though there were peaks and valleys, that number steadily increased to nearly 10 million in March 2006. The number dipped slightly but remained steady at around 9 million through October, the last month for which Big Champagne has data.")

2. "2006 10-Year Music Consumer Trends Chart," RIAA, available at link #111 (last visited July 31, 2007).

3. Lydia Pallas Loren, "Building a Reliable Semicommons of Creative Works, Enforcement of Creative Commons Licenses and Limited Abandonment of Copyright," *George Mason Law Review* 14 (2007), available at link #112.

4. Mia Garlick, "Lonely Island," Creative Commons, available at link #113.

5. von Hippel, *Democratizing Innovation,* 91.

6. Ibid., 78–79.

7. Ibid., 86.

Routine and intentional free revealing among profit-seeking firms was first described by Allen (1983). He noticed the phenomenon, which he called collective invention, in historical records from the nineteenth-century English iron industry. In that industry, ore was processed into iron by means of large furnaces heated to very high temperatures. Two attributes of the furnaces used had been steadily improved during the period 1850–1875: chimney height had been increased and the temperature of the combustion air pumped into the furnace during operation had been raised. These two technical changes significantly and progressively improved the energy efficiency of iron production—a very important matter for producers. Allen noted the surprising fact that employees of competing firms publicly revealed information on their furnace design improvements and related performance data in meetings of professional societies and in published material. (Ibid., 78.)

8. Ibid., 80.

9. William J. Baumol, *The Free-Market Innovation Machine: Analyzing the Growth Miracle of Capitalism* (Princeton, N.J.: Princeton University Press, 2002), 134–35. See also Mark Lemley and Brett Frischmann, "Spillovers," *Columbia Law Review* 100 (2006).

10. von Hippel, *Democratizing Innovation*, 87.

11. Baumol, *The Free-Market Innovation Machine*, 120.

12. von Hippel, *Democratizing Innovation*, 42.

13. Ibid., 23.

14. Ibid., 22.

15. Paul Krill, "Sun: Pay Open-Source Developers," *InfoWorld*, May 7, 2007, available at link #114.

16. Tapscott and Williams, *Wikinomics*, 206–7.

17. Ibid., 205.

18. Ibid., 206.

19. Ibid., 209.

20. Ibid., 210.

21. All quotes from David Marglin taken from an interview conducted June 12, 2007, by telephone.

22. Robert Lemos, "Companies Fight over CD Listings, Leaving the Public Behind," CNET News.com, May 24, 2001, available at link #115.

23. Tim Deal, *User-Generated Video on the Web: A Taxonomy and Market Outlook* (Silver Spring, Md.: Pike & Fischer, 2007), 11.

24. See Richard Stallman, "The GNU Manifesto," GNU Project, available at link #116 (last visited August 20, 2007). ("I consider that the golden rule requires that if I like a program I must share it with other people who like it.")

25. See Lessig, *The Future of Ideas*, 70.

26. "Official Rules for ACIDplanet.com, David Bowie Remix," ACIDplanet.com, available at link #117 (last visited July 31, 2007).

27. Elizabeth Durack, "fans.starwars.con," Echo Station, available at link #118 (last visited July 31, 2007).

Chapter 9. Reforming Law

1. See 17 U.S.C. §§108, 112, 403, 512, 1201, 1203, 1204, 1309.

2. See, e.g., 17 U.S.C. §115 (establishing a compulsory license for making and distributing phonorecords).

3. Lawrence Lessig, "The Regulation of Social Meaning," *University of Chicago Law Review* 62 (1995), 943–1045.

4. "Googling Copyrights," *Wall Street Journal*, October 3, 2005.

5. See Brian Lavoie, Lynn Silipigni Connaway, and Lorcan Dempsey, "Anatomy of Aggregate Collections: The Example of Google Print for Libraries," *D-Lib Magazine*, September 2005, available at link #119.

6. See the data in Paul J. Heald, "Property Rights and the Efficient Exploitation of Copyrighted Works: An Empirical Analysis of Public Domain and Copyrighted Fiction Best Sellers" (January 9, 2007), UGA Legal Studies Research Paper No. 07-003, available at link #120.

7. Richard Epstein, *Simple Rules for a Complex World* (Cambridge, Mass.: Harvard University Press, 1995).

8. R. Anthony Reese, "Innocent Infringement in U.S. Copyright Law: A History," *Columbia Journal of Law & the Arts* 30 (2007), 133–84.

9. As Patterson explains, before 1909, the law included the word "copies," but in a section defining the scope of the rights, the law made clear that the exclusive right to "copies" did not apply to a "book." Instead, the right was intended to protect works, such as statues, that could only be "copied." L. Ray Patterson, "Free Speech, Copyright, and Fair Use," *Vanderbilt Law Review* 40 (1987): 40–43.

10. Ibid.

11. Jessica Litman, "The Exclusive Right to Read," *Cardozo Arts & Entertainment Law Journal* 13 (1994): 29, 34–35.

12. There are of course important limits on Congress's power if it is to live up to the obligations of international law. I don't address those limits here. The simplest way to avoid inconsistency yet permit significant reform would be to limit the reach of any reform to U.S. law alone. More ambitiously, the United States could take the lead in reforming international law to make it conform better to creative interests. Christopher Sprigman, "Reform(aliz)ing Copyright," *Stanford Law Review* 57 (2004): 485.

13. See William W. Fisher, *Promises to Keep* (Stanford, Calif.: Stanford University Press, 2004); Neil Weinstock Netanel, "Impose a Non-commercial Use Levy to Allow Free P2P File-sharing," *Harvard Journal of Law and Technology* 17 (2003): 1; "A Better Way Forward: Voluntary Collective Licensing of Music File Sharing," Electronic Frontier Foundation, available at link #121 (last visited January 18, 2008).

Chapter 10. Reforming Us

1. David Hackett Fischer, *Albion's Seed: Four British Folkways in America* (New York: Oxford University Press, 1989), 765.

2. Lessig, *The Future of Ideas,* 4.

3. Jenkins, *Convergence Culture,* 134.

4. Ibid., 18.

5. James C. Carter, *The Provinces of the Written and the Unwritten Law* (New York: Banks & Brothers, 1889), 4.

6. James C. Carter, *Law: Its Origin, Growth, and Function* (New York: G. P. Putnam's Sons, 1907), 323.

7. My favorites are Lawrence Wright, *The Looming Tower: Al-Qaeda and the Road to 9/11* (New York: Knopf, 2006) and Bob Woodward, *State of Denial: Bush at War, Part III* (New York: Simon & Schuster, 2006).

8. The simplest claim to support here is that if kids view laws regulating culture as unjust, they are less likely to obey those laws. As Professor Geraldine Moohr argues, "a criminal law that is not supported by community consensus will be less effective and can even be counterproductive. Members of the community will not condemn those who violate such laws. This state of affairs can eventually weaken respect for the law. Witnessing punishment for conduct not viewed as immoral may cause people to view the law as less than legitimate and not morally credible. For this reason, courts have been generally cautious when deciding whether conduct in which citizens routinely engage is a crime." Geral-

dine S. Moohr, "The Crime of Copyright Infringement: An Inquiry Based on Morality, Harm, and Criminal Theory," *Boston University Law Review* 83 (2003): 731, citing Paul H. Robinson and John M. Darley, *Justice, Liability and Blame* (Boulder, Colo.: Westview Press, 1995). As Moohr concludes, "criminalizing copyright infringement may produce the opposite of its intended goal." Similar conclusions have been reached studying other "youth crimes," such as illegal use of alcohol, tobacco, and marijuana. See Claudia Amonini and Robert J. Donavan, "The Relationship Between Youth's Moral and Legal Perceptions of Alcohol, Tobacco and Marijuana and Use of These Substances," *Health Education Research* (2005): 276.

The harder claim to sustain is that any effect localized around culture crimes might bleed to other areas of the law. Scott Menard and David Huizinga have advanced an important understanding about the interaction between conventional attitudes and delinquent behavior in adolescence, suggesting that changes in attitudes can lead to a small increase in delinquent behavior, which in turn will have a reinforcing effect on attitudes. See Scott Menard and David Hulzing, "Changes in Conventional Attitudes: and Delinquent Behavior in Adolescence," *Youth and Society* 26 (1994): 23. But the most important, and foundational work supporting this hypothesis is Tom Tyler's *Why People Obey the Law* (New Haven, Conn.: Yale University Press, 1990), 161. Much great work has been built upon Tyler's foundation. But the core insight Tyler advanced in this debate—that the "values that lead people to comply voluntarily with legal rules" ... "form the basis for the effective functioning of legal authorities"—underlies the concern that local skepticism (or even disgust) with criminal enforcement of laws regulating behavior perceived to be harmless might generalize beyond that locality. Tyler's particular concern was procedural legitimacy. The *in terrorem* tactics of the RIAA and MPAA certainly weaken any perceived procedural legitimacy to the enforcement of these culture crimes.

The strongest support for the idea that perceived injustice in one law can spill over to others comes from the extraordinary work of Professor Janice Nadler. In her essay "Flouting the Law," *Texas Law Review* 83 (2005): 1399, she provides experimental evident to support the hypothesis that willingness to disobey can extend far beyond a particular unjust law.

The most obvious or appealing parallel—to youth in the Soviet Union—is a harder claim to sustain. Paradoxically, even though youth in the Soviet Union were referred to as the "bewildered generation"—bewildered by the hypocrisy and double standards of the late Soviet Union especially—"the Soviet Union was unique in its attempt to control and shape the development of its youth into 'proper Communist citizens'" (James O. Finckenauer, *Russian Youth* [New Brunswick; N.J.: Transaction Publishers, 1995], 80). See also June Louin-Tapp, "The Geography of Legal Socialization," *Droit Et Société* 19 (1991): 331, 349 ("Soviet youth perceive USSR law and its applications to be more fair than American youth perceive US law and its applications"). Thus, while there's little doubt that juvenile crime was rising by the end of the Soviet Union, it is difficult to compare Soviet attitudes with American attitudes. Both may suffer the same negative effect (laws seen to be unjust), but only one had an extensive propaganda effort to counter the consequences of that effect (the Soviet Union). See also Walter D. Connor, "Juvenile Delinquency in the USSR," *American Sociological Review* 35 (1970): 283 (concluding delinquency not "protest"). See also Emanuela Carbonara, Francesco Parisi, and Georg von Wangenheim, "Unjust Laws and Illegal Normas," *Minnesota Legal Studies Research Paper* No. 08–03 (January 2008) (modeling effect of social opposition to unjust laws on effects of legal intervention).

9. Anonymous, "Who Passes Up the Free Lunch" (unpublished, 2007) (on file with author).

Conclusion

1. See Carl J. Dahlman, "The Problem of Externality," *Journal of Law and Economics* 22 (1979): 141.

2. Thomas Jefferson letter to Isaac Mcpherson, August 13, 1813, reprinted in H. A. Washington, ed., *Writings of Thomas Jefferson 1790–1826,* vol. 6 (Washington, D. C: Taylor & Maury, 1854), 180–81; quoted in *Graham v. John Deere Company of Kansas,* 383 U. S. 1, 8–9n.2 (1966).

3. See Mark Stefik, ed., "Epilogue: Choices and Dreams," *Internet Dreams: Archetypes, Myths, and Metaphors* (Cambridge, Mass.: MIT Press, 1996), 391.

4. Rick E. Bruner, "Blogging Is Booming," *iMedia Connection,* April 5, 2004, available at link #122 (last visited January 18, 2008).

5. Stephen Breyer, "The Uneasy Case for Copyright: A Study of Copyright in Books, Photocopies, and Computer Programs," *Harvard Law Review* 84 (1970): 281.

6. See Lessig, *Code Version 2.0,* 409n8.

7. Al Gore, *The Assault on Reason* (New York: Penguin Press, 2007), 194.

8. Ibid., 195.

INDEX